Athabasca
OIL SANDS

This publication was funded in part by a generous endowment from Syncrude Canada Ltd.

Athabasca
OIL SANDS
Northern Resource Exploration, 1875-1951

By Barry Glen Ferguson

Alberta Culture/Canadian Plains Research Center

1985

Canadian Cataloguing in Publication Data

Ferguson, Barry Glen, 1952-
    Athabasca oil sands

    Co-published by Alberta Culture.
    Bibliography: p.
    Includes index.
    ISBN 0-88977-039-5

    1. Athabasca tar sands (Alta.) — History.
2. Oil sands industry — Alberta — History.
I. University of Regina. Canadian Plains
Research Centre.   II. Alberta. Alberta
Culture.   III. Title.
TN873.C22A443 1985   338.2'7282'0971232
                             C85-091370-5

Contents

List of Maps

Acknowledgements

Archivists at all the libraries and archives I worked in were extremely co-operative. In particular, James Parker at the University of Alberta Archives was most patient, and his own historical work on the Athabasca region made his advice doubly important. David Leonard at the Provincial Archives of Alberta went to extraordinary lengths to find material for my use. At the Public Archives of Canada, Public Records Division, Terry Cook, Robert Hayward and David Smith were most helpful and saved me a great deal of time.

A number of historians have discussed this work with me and have shared information. These are John S. Nicks, Donald Wetherell and Doug Owram in Edmonton, and David Finch, Douglas Cass, Howard Palmer and Anthony W. Rasporich in Calgary.

Alberta Culture's historians were crucial associates. Frits Pannekoek, director of the Historic Sites Service, encouraged the original project; Ian Clarke and Leslie Hurt endured more difficulties than they deserved in administering it; Carl Betke has shown insight as the manuscript has been guided through the evaluation and editing processes assisted by Judy Larmour.

Kent Oliver, head of petroleum technology at the Northern Alberta Institute of Technology, consulted with me and worked hard on the study of oil sands technology which is contained in this book's Appendix.

Canstar Oil Sands Ltd. generously provided flights into the Bitumount site from Fort McMurray on several occasions in 1981.

Sharon Mackenzie typed vast portions of the manuscript quickly and efficiently.

My anonymous reviewers mixed criticism, flattery and considerable knowledge in their assessments of the manuscript; their suggestions were most helpful.

Gillian Wadsworth Minifie, Co-ordinator of Publications at the Canadian Plains Research Center, University of Regina, brought order and clarity to the manuscript and was most genial throughout.

Natalie Johnson has put up with me while I worked on the manuscript and provided much assistance throughout.

I want to make it clear that the Historic Sites Service of Alberta Culture, which funded the original study and helped publish the present work, has allowed me total freedom in developing my arguments.

The responsibility for this work, for better or worse, is entirely my own.

Introduction

This work has its origins in an historical study for Alberta Culture in which I undertook "to report on the history of oil sands development in the Athabasca River Bituminous Sands region, with an emphasis on the Bitumount Provincial Historic Resource." The study which resulted from this undertaking is based on available primary sources, mostly unpublished government records. It examines the ways in which men have understood the importance of the bituminous or oil sands and surveys their attempt to exploit the sands from the 1870s to the 1950s.

The Bitumount Historic Resource is the partly intact site of a Province of Alberta oil sands mining, separating and refining operation, completed in 1948, which was the last and most successful of the many engineering research facilities constructed before commercial development. The Bitumount site and its record of activity comprise the culmination of this period of pioneering engineering and exploratory activity, at its most intensive during the 1920s, 1930s, and 1940s. All this occurred prior to the commitment and shift to effective commercial development of the oil sands as a major petroleum resource, which began in the 1950s.

My study focusses on this pioneering era, from the earliest scientific work in the 1870s through various federal government, provincial government, and private-sponsored research and experimental activities to the early 1950s, when the Bitumount plant operations were subjected to serious study and the consensus was reached that commercial exploitation should occur. The "history of oil sands development" that I have studied ends in 1951. The subsequent era, one in which the existing mammoth operations at Suncor and Syncrude were built and operated, is not one for which private or government records are available; no serious historical study will be possible until such records are opened.

It is fitting for three reasons to study the oil sands history in such a way that the Bitumount plant is both its focus and conclusion. First, Bitumount's completion and results comprised a critical turning point in the history of the oil sands. The processes improved at Bitumount were those which had emerged as a result of thirty years of hard work and often-failed research by government and private experimenters. Moreover, with Bitumount, experimental work was yielding to commercial planning. This leads to a second point: Bitumount is a monument to the processes which have been developed into a major industry and source of petroleum in Canada. Bitumount contains a simple version of the plants at Fort McMurray which currently exploit the oil sands. Third, as mentioned above, serious study of the oil sands can at present be conducted only up to 1951, when access to proper records ends. (I exclude newspapers as a source of historical information, save as a testament to opinion and a measure of uninformed speculation.)

* * *

The bituminous cliffs along the Athabasca River are shown in this photograph taken by pioneer oil sands researcher Sidney Blair in the 1920s.

The Athabasca tar sands, located in north-central Alberta, are a huge deposit of bitumen or crude petroleum mixed with sand and other mineral materials. The deposits straddle the Athabasca River between its confluences with the Clearwater and Firebag Rivers |see Map 1|. The deposits vary in concentration of bitumen and in the amount of overburden which seals the resource; along portions of the Athabasca River, the oil sands are exposed, while in other areas they are buried by hundreds of feet of gravel and earth.

Before beginning the story of the oil sands as a scientific and commercial project, the extent, the characteristics, and the value of the oil sands should be examined.

The phrases "tar sands" and "oil sands" are commonly used to describe the Athabasca deposits and they are used throughout this book. The terms are evocative and accurate enough. The "tar sands" are a reservoir of bitumen, the generic word for liquid hydrocarbons. The particular bitumen contained in the Athabasca sands is a tar-like substance, a mixture of hydrocarbons notable for being very thick at ordinary temperatures and

University of Alberta Archives

5

An aerial view of the Bitumount separation plant, the refinery, and the camp, circa 1949.   Provincial Archives of Alberta

containing higher carbon and sulphur content than ordinary or conventional crude oil. In comparison, conventional oils are referred to as "light" crudes, in reference to their viscosity, colour and their greater hydrogen content. Engineer Donald Towson's 1983 definition of the "tar sands" captures these important factors, noting also that the bitumen in tar sands is "not recoverable in its natural state through a well by ordinary production methods." Towson also discusses the important factor that bitumen from the tar sands is "hydrogen-deficient oil" which requires either "carbon removal (coking) or hydrogen addition (hydro-cracking)" to be used. In other words, bitumen from the tar sands is not as ready for refining as most conventional oil from wells; hence the term "synthetic" oil often used for the refined product from the tar sands. The current legal definition of the bituminous sands, comprising bitumen found in a geographical area in which oil sands are found, is based on the conception of the deposits as a reservoir of crude bitumen or crude oil contained in mixtures of sand or other rock materials and not recoverable by conventional wells.[1]

Map 1.   Alberta Oil Sands Deposits in Relation to Alberta and
          Canada.

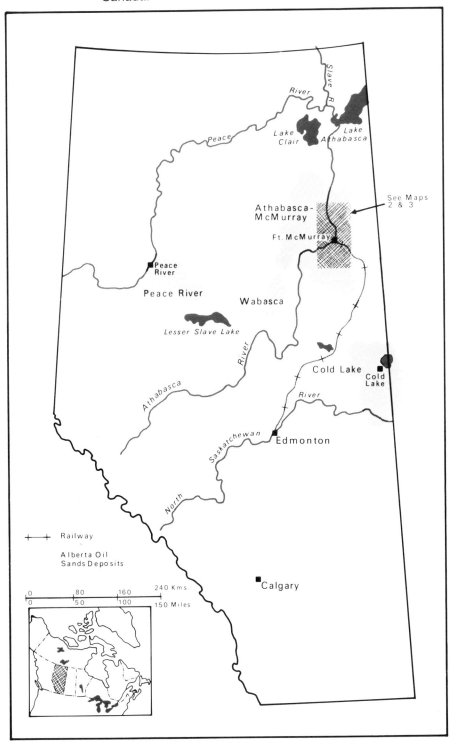

The characteristics of the bituminous sands suggest the difficulties which their development presented. Three general problems required the attention of researchers and developers. First, the physical and chemical analyses of the resource had to be made. Second, processes had to be found which would allow the separation of bitumen from the oil sands. Third, commercial uses had to be assessed. The Athabasca tar sands are located in a remote corner of the province that was and is removed from concentrated markets and centres of finance and expertise (see Map 1).

Alberta's tar sands along the Athabasca are the largest by far of four deposits in north-central Alberta. The Athabasca area contains some three-quarters of the total oil sands reserves in Alberta and most of the recoverable reserves. Moreover, the Athabasca tar sands comprise one of the two major deposits of oil sands in the world. There are known to be 16 major tar sands deposits containing some 2,100 billion barrels in potential petroleum. Of the 7 largest, Alberta's Athabasca River and Venezuela's Orinoco River deposits are the two largest—the elephant pools, to use the petroleum term for huge deposits. The Orinoco deposits contain over 1,000 billion barrels while the Athabasca reserves hold about 800 billion barrels of crude oil in total. There is also a large deposit (estimated 500 million barrels in place) on Melville Island in the Canadian Arctic.[2]

To gain some perspective on the value of the Alberta oil sands, a couple of points might be made. First, the total reserves do not equal the amount of oil that may be recovered given present engineering skills and present oil prices. In the case of the Athabasca reserves alone, estimates of recoverable reserves vary somewhat, but it is now usually suggested that only about 10 percent of the reserves are mineable (and mining is the basis for current oil sands extraction). Further, most estimates indicate that about one-half of the total reserves may be recoverable if major engineering advances occur. These advances would require the perfection of so-called *in situ* or on-site recovery of the bitumen by drilling and pumping operations. Most of the research of the past twenty years has been pointed toward *in situ* processing. The result of these limitations, given that the concentrations of oil sands vary so greatly, is that exploitation is limited to areas that contain above 8 to 10 percent oil in mineral matter and in areas that have less than 150 feet of overburden. An added concern is that about one-third of the mineable oil sands contain quantities of bitumen that are recoverable at the present time. All this means that about 75 billion barrels of oil are mineable but only some 25 billion barrels of oil are assuredly and currently recoverable in the Athabasca deposits.[3]

The second aspect of Alberta's oil sands reserves is their size compared to Canada's and world reserves. This sort of estimate has become notoriously variable in recent years, certainly since the first upward shift of world oil prices so dramatically affected supply and pricing beginning in the autumn of 1973.

Variability of supply and price have left Canada in a rather ambiguous position, however. Alberta contains comparatively large reserves of conventional oil as well as the tar sands. More recently, the Canada Lands areas of the Arctic and Atlantic oceans appear to offer considerable potential as well, recently estimated at 2.6 billion barrels according to the Geological Survey of Canada. In any case, the Alberta government estimates that the province has some 4.5 billion barrels of conventional oil reserves. While these Alberta reserves could easily be improved through future discoveries or technological changes in recovery processes, at present they comprise a total that is about one-fifth the size of current recoverable oil sands reserves.[4]

The Athabasca deposits, then, contain overwhelmingly important reserves for Alberta and Canada. But these reserves are not at present the petroleum cornucopia that they have often been made out to be. For instance, the current recoverable oil sands reserves are far smaller than existing conventional reserves in the Middle East, which are at least eight times their size. Even current total North American reserves, including Mexico's considerable fields, are more than twice the size of the proven recoverable oil sands reserves.[5] This leads to reiteration of the telling point that the existing limits on recovery do continue to prevent the tapping of the oil cornucopia that recovery of the unmineable portions of the oil sands indeed would bring about. If these limits were breached, the Athabasca (and Orinoco) reserves would provide petroleum that would guarantee world supply for many, many decades.

The limits to current recovery and the alluring potential of future reserves together explain the continuing and considerable efforts to develop new techniques of extracting oil from the oil sands. The amount of research undertaken by government agencies alone is significant. The government of Alberta has created a research agency, Alberta Oil Sands Technology and Research Authority (AOSTRA), which has co-ordinated and supported public and much private investigation since 1975.

AOSTRA spent more than $100 million in a five-year programme funded jointly by the federal government. In 1980, AOSTRA spent $29.2 million; in comparison, the entire budget of the Research Council of Alberta was $22.5 million. Since 1980, AOSTRA's budget has increased; in 1984, it committed more than $37.7 million to tar sands research. A number of AOSTRA projects are joint undertakings with private oil companies. These firms have conducted and are conducting their own research at a number of sites in the oil sands areas. As of March 1985, 17 projects were underway in the Athabasca sands, 5 with AOSTRA participation. Provincially-financed research indicates the continuing role of government in furthering technical development and solving the continuing problem of finding a means to ensure efficient and commercial production from the 90 percent of oil sands not accessible by mining.[6]

The current level of research suggests the oil sands are a technical novelty when, in fact, they are not. The following pages will show the long commitment of government and business to resolve the problems of oil sands exploration.

I: The Dominion Government: Assessing the Tar Sands 1875-1918

The bituminous sands have been known to commercially-minded men since fur traders began pressing into the Athabasca-Clearwater rivers system in the late eighteenth century. But the shift from anecdotal knowledge to scientific assessment began with the involvement of the government of Canada's geological and geographical investigations of the 1870s and 1880s. The impetus for serious exploitation of the tar sands came with the government's reassessment of the importance of petroleum as a resource, beginning about 1912.

* * *

The Athabasca country was opened to commerce by Peter Pond, a semi-literate, ruthless Yankee fur trader. Pond explored the fringes of the Arctic Ocean drainage system in the late eighteenth century in his search for more and newer sources of animal pelts. His reconnaissance of the Clearwater and Athabasca Rivers in 1778 led to the exploitation of the area as a rich fur-trading region from the 1780s. The region was, therefore, subject to numerous commentaries by fur traders and other explorers.[1]

The development of the fur trade and its expansion north and west into the Mackenzie-Peace rivers system meant that there was considerable commercial activity in the area. The Methy Portage route between the Churchill and Clearwater rivers meant that the Saskatchewan and Mackenzie drainage—and fur trade—systems were connected. British and Canadian intrusion upon the area and the ways of the Indian inhabitants was also continuous. The tar sands were known as a curiosity which had some limited uses as a pitch for canoe repairs, or so Alexander Mackenzie reported in the 1790s. In his reflections on the fur trade, Mackenzie described the "bituminous fountains" or pools he found on the Athabasca. These "fountains" were collections of bitumen

> into which a pole of twenty feet long may be inserted without the least resistance. The bitumen is in a fluid state, and when mixed with gum, or the resinous substance collected from the spruce fir, serves to gum the canoes. In its heated state it emits a smell like that of sea-coal. The banks of the river, which are there very elevated, discover veins of the same bituminous quality.

Impressed by the profuse amounts of liquid bitumen, Mackenzie also noted that the bitumen was easily gathered from natural collecting spots. This fact also impressed later observers, easily leading to the conclusion that the tar sands merely rested atop pure reservoirs of bitumen.[2] This was the conclusion of the Englishman, John Richardson, in 1848. He toured the Athabasca country, where he found a "mineral pitch" made of sand and bitumen. Richardson noted that the "country is so full of bitumen that it flows readily into a pit dug a few feet below the surface."

Unlike the explorer Mackenzie, however, the tourist Richardson did not notice any commercial uses or potential for the bituminous sands.[3]

What this activity and the fur traders' commentaries indicate is simply that the oil sands were known and occasionally assessed by visitors. But the nature of the deposits and their possible commercial uses remained open questions until the late nineteenth century when Canadian government scientists began to make their systematic inventory of the nation's resources. This occurred most strenuously after 1870 when Canada acquired the Hudson's Bay Company lands in the western interior of the continent.[4]

When Canada took title to the 2.5 million square miles of the continental interior, the government began to approach the considerable problems of administering and developing the area for settlement and commerce. Among the many steps taken (if the mixture of stridency, befuddlement, bravado and vision that characterized government and individual effort in response to the complexity of the problems in taking over the area can be so explained) was the inventory of Canada's physical resources. This inventory was the task of the Geological Survey of Canada. The Survey had been formed by the colonial legislature of the Province of Canada in 1842 in order to provide the accurate knowledge that would encourage settlement and resource development. The remarkable explorers and scientists of the Survey undertook studies of the topology, the geology, the people, as well as fauna and flora of the colony and later the nation, in a systematic and highly utilitarian manner. After 1870, the government charged the Geological Survey with providing the survey of the Canadian-American boundary and with assessing accurately the resource and agricultural potential of the new North-West Territories.[5]

One small aspect of the Survey's prodigious work was the careful scrutiny of the land and resources in the Athabasca River region. The Survey's work here, as in all of its endeavours, provided the crucial first analysis of the Athabasca oil sands. It also marks the initial example of the vital role government enterprise had in assessing and encouraging commercial interest in the oil sands.[6]

The first Geological Survey expedition to assess the Athabasca region was led by a distinguished botanist and western explorer, John Macoun. Macoun was largely self-taught but he was knowledgeable enough to teach at Albert College in Belleville (later part of the University of Toronto) and then to work full-time for the Geological Survey, while writing a great deal about the Canadian prairies. Macoun was part of two expeditions to the interior in the early 1870s. One of these was the famous Fleming-Grant survey of a Pacific Railway route in 1872-73; the other was a Geological Survey examination of the rivers draining into Lake Athabasca in 1875. Macoun's description of the "oil shales" and "tar sands" of the Athabasca River was gleaned as he canoed from the Lower Peace River through Lake Athabasca and up the Athabasca River. Macoun was a highly optimistic supporter of prairie and northern settlement; he described the Peace-Athabasca delta country as "very Manitoba in embryo ...." The botanist, however, was less certain about the prospects of the Athabasca country as he ascended the river.[7]

Macoun was a vivid reporter and he recorded the geographical and botanical features he observed. He described the occurrences of "tar sand" as being visible along the Athabasca River from the Embarras upstream to the Clearwater, where, he noted, the tar beds disappeared beneath the overburden. Macoun's description of the river shore is worth recalling, for it remains evocative of the physical presence of the oil sands on the Athabasca's banks. "I found," he wrote,

> a light grey sandstone, partly saturated with the tar, and overlaying this there was at least fifteen feet of it completely saturated, and over this again, shale largely charged with alkaline matter.

Macoun also described the physical properties of the "tar" itself:

> When we landed, the ooze from the bank had flowed down the slope into the water and formed a tarred surface extending along the beach over one hundred yards, and as hard as iron; but in bright sunshine the surface is quite soft, and the men tracking along shore often sink into it up to their ankles.

He left the impression that an eminently available resource was present. His key observation was that the tar was apparently flowing out of the sand and rock rather than remaining mixed with other matter.[8]

In a later passage in his account, Macoun also observed another agglomeration of "pure" tar. His party surveyed the plateau surrounding the Athabasca River where Macoun came across a small creek or spring which caught his attention. He found that beneath the water's surface there was a layer of what he termed "pure tar". He went on:

> I noticed a little stream of water flowing into the pool, which was coated with an oily scum, and under the stream an abundance of tar. Along the beach it was seen oozing out in many places, and by gathering and washing the sand saturated with it, we obtained just as pure tar as we brought from the spring on the hillside.[9]

Macoun again made the point that tar was oozing from the earth. He also made a more prescient if inadvertent point. Macoun described both the spring's tarry bed and the river shore's tar deposits as containing "pure tar." He witnessed a natural washing out of the sand which was mixed with the bitumen or tar. In this he observed a natural version of the water-washing that is the essence of man's technology for extracting bitumen from oil sands. Of course, Macoun's observation was simply a keen amateur's, but he had seen an important phenomenon of the behaviour of the oil sands. But Macoun's main impression— it can hardly be called a conclusion—was that the "tar" was not mixed with mineral matter; rather the tar was flowing through it.

Macoun's work was only a preliminary foray and in 1882 Robert Bell led an expedition to study the Athabasca Basin. Bell was a major figure in the Survey's work for over thirty years. He was a Canadian-born geologist, trained by his field work and study at Queen's and Edinburgh universities. His work on the geography and geology of northern Canada made a great contribution to knowledge of the country, although his career was marred by considerable controversy due to his frustrated ambition to head the Survey (he was acting director from 1901 to 1906).[10]

Bell's 1882 trip into the Athabasca involved a 2,000-mile circuit of the Athabasca country and a 5,600-mile round trip from Ottawa, a rather typical summer outing for a Geological Survey party. Bell went over the Athabasca and its tributaries and then composed his assessment of the value of the many mineral resources he found, as well as of the geological formation of the region. Bell noted the physical impression and characteristics of the "asphaltic sands" in a fashion similar to Macoun. He wrote of the tendency to plasticity of the "sandy pitch" when temperature rose and the occasional concentrations of "pools of oil and then tar."[11]

Bell moved beyond description, for he explained his views of just how the asphaltic sands occurred and what uses they might have. He followed correct geological thinking in assuming that the "vast quantities of somewhat altered petroleum" in the sands had arisen from the Devonian limestones which underlie the Athabasca area. He also argued that the asphalts or petroleums had indeed flowed out of the limestone into the sandstone layers at or just beneath the surface of the region. He explained why this was so:

> *That the petroleum came from below would be expected in accordance with natural laws, and from the fact that the higher rocks of these regions to the south and west would have been very unlikely to produce any petroleum, even if they had once extended all over this region.*

Bell thought that the clay materials in the surface layers prevented the continued out-flow of "petroleum," as he called the bitumen, though in some places he thought that no such inhibition was present. Drilling would uncover "isolated springs or wells" of petroleum.[12]

His argument about the likely origins of the oil in the sands led Bell to an obvious conclusion:

> *The enormous quantity of asphalt, or thickened petroleum, in such a depth and extent of sand indicates an abundant origin. It is hardly likely that the source from whence it came is exhausted. The whole of the liquid petroleum may have escaped in some parts of the area below the sandstone, while in others it is probably still imprisoned in great quantities and may be found by boring.*

Bell thus announced a theory about the petroleum resource and its geological condition that would guide research for twenty years at least. The oil sands rested upon larger reservoirs of petroleum. The hopefulness of Bell's deductions led him to somewhat speculative conclusions about the commercial potential of the oil sands, although it should be noted that he did not indulge in any extravagant claims about immediate development. Bell's general geological observations about the tar sands stand up well to later work (as does Bell's work in general), and it would be unfair to scoff at his erroneous view about the presence of "pools" of bitumen in the limestone. Based upon sound judgement and study, then, Bell announced that the Athabasca oil sands presaged discoveries of important petroleum deposits.[13]

Bell also had insinuated into his account the assurance that the tar sands contained a valuable *petroleum* resource. This was a fact of some importance, too, for it made clear during what can now be seen as the early stage of the petroleum age of economic development that the Athabasca deposits contained a product of more than passing usefulness.

16

This 1924 photograph of a bituminous sands cut on the Christina River shows a scene similar to some Robert Bell might have encountered during his 1882 survey trip along the Athabasca and its tributaries.

Bell was able to assure his readers that the tar sands contained an asphaltic base because his Ottawa colleague, G. Christian Hoffmann, had conducted chemical tests on the oil sands. Hoffmann was an Australian who for many years acted as chemist and metallurgist for the Survey, conducting most of his work on the many mineral samples collected by Survey geologists. Bell provided Hoffmann with samples even before his major expedition. Hoffmann himself described the tests, noting that alcohol did not dissolve bitumen from the sands. But he observed that petroleums like kerosene and benzine did dissolve the bitumen, leaving what he termed "a pure or almost pure siliceous residue...."[14] Hoffmann did not comment on the commercial potential of the oil sands or the separation process in his 1880-81 work.

But Bell did elaborate on these themes. First he noted that Hoffmann and two other researchers, Isaac Waterman, a London, Ontario, refiner, and Lieutenant Cochrane of the chemistry department at the Royal Military College, Kingston, had confirmed the bitumen content of the sand. Indeed, they estimated that content at between 12 and 15 percent of the volume of oil sands. Bell wrote that Hoffmann had also attempted to separate the bitumen from the sand in a more novel process than he had reported. Hoffmann managed to extract "69.26 percent of the bitumen" in the sands by "boiling or macerating" oil sands in water. The bitumen, however, contained 50.1 percent sand.[15] This prescient experiment sustained Bell's view of the ultimate importance of the oil sands. Neither Bell nor Hoffmann stated whether the experiment in separation was attempted because there were doubts about the existence of pools of bitumen. Nor did either scientist elaborate on the pioneering effort at hot water separation.

To support his claim that the oil sands constituted a resource certain to have commercial value, Bell made a number of points. The Hudson's Bay Company had collected "pitch" from the concentration of bitumen that existed along shore and at inland springs. This pitch was barrelled and used for boat and, Bell claimed, roof repairs throughout the Hudson's Bay Company's operations. (In this regard—and however extensive it was remains unclear—the Baymen were continuing a practice which had been carried out, at least to a limited extent, by the native

Indians of the region.) The fact that the fur traders could distribute the pitch for various uses served as a model for Bell's plans for the more modern exploitation of the oil sands. Any problem of transporting the bitumen found on the Athabasca Bell easily explained. He noted that the extensive water transportation network and the soon-to-be-built Canadian Pacific Railway (CPR) each provided effective ways of shipping petroleums to markets.[16]

Although Bell's estimation about the resource may now seem too hopeful, it was sensible enough. In fact, the 1880s and 1890s comprised a period in which both present and future economic well-being were seen in such developments as the completion of the CPR, the initial exploitation of the huge Sudbury Basin mineral deposits and the many surveys of western agriculture. In 1887-88, a Senate committee on the "Great Mackenzie Basin" surveyed with great hope the agricultural and mineral resources of the entire Mackenzie River drainage system. The Peace River's agricultural potential, the petroleum possibilities, the Mackenzie's own farming resources, each indicated that exploitation was imminent.[17]

This optimism was not foolish, for it was based on available scientific information and on the maturing political economic plans of the Canadian government. Optimism, of course, was somewhat premature. But it explains the next venture in the investigation of the tar sands. Accordingly, and typically, further government-sponsored research was called for to clinch the case for development so that private enterprise might take over.

Further study of the resource was conducted by the Geological Survey's Robert McConnell. McConnell was a McGill graduate who earned a high reputation as an explorer and administrator in the Survey; he eventually became deputy minister of mines. He tried to estimate the actual extent and value of the bitumen deposits, noting that at least 1,000 square miles of tar sands existed, with a thickness varying from 150 to 225 feet. Since the average bitumen content was about 12 percent, McConnell noted from Hoffmann's analysis, he calculated that there were 6.5 cubic miles of bitumen or 30 million tons in the sand. This emphasis on the resource contained in the oil sands indicates a slight shift away from Bell's argument about the likelihood of pools of bitumen being discovered. McConnell in 1890 wrote about the value of the "tar sands themselves," although he continued to maintain Bell's point that pools must still exist in parts of the region. If this shift in perspective indicates a degree of uncertainty about the sand and the bitumen that Bell had not allowed for, it did not seem that the Geological Survey was abandoning its responsibility to carry out further work.[18]

The solution to the perplexity he felt was only to be gained by drilling in the area, McConnell concluded. He proposed that core holes should be drilled at a couple of places in the Athabasca country. This action would establish the possible southern limits to the oil sands and "settle" the question of oil pools that McConnell clearly thought needed answering.[19]

Accordingly, in 1894, an initial $7,000 was voted by Parliament for experimental drilling. The survey appointed A.W. Fraser, an experienced petroleum driller from Ontario, to conduct the work. Fraser began by setting up a drilling rig at Athabasca Landing. His crew drilled to a depth of 1,100 feet by October 1894, an impressive achievement for the cable-tool rigs of the time. Fraser's crew drilled through two pockets of natural gas, at least one of which was a problem for them. But the results were inconclusive in establishing the existence of bitumen pools. Fraser concluded that the discovery of natural gas was very heartening, that commercial activity should be stimulated and that the limits of the tar sands deposits were easy to determine.[20]

The first Athabasca well was abandoned after it reached 1,500 feet, which occurred only after two further summers of work. In 1897, two other wells were planned. One was at the confluence of the Pelican and the Athabasca and the other at the Victoria Settlement (a Methodist mission) on the North Saskatchewan River. The Pelican well drilled through a layer of tar sand at 750 feet but hit a large gas reservoir at 820 feet immediately below the tar sand layer. The gas was allowed to escape to enable drilling the next year. Still, the results harbored one major disappointment. "Maltha or natural tar" was the only substance found in the rocks and no "petroleum proper" or unadulterated pools of bitumen were discovered. This suggested that at Pelican river, at least, there were no pools of bitumen. The site, however, is at the heart of a tar sands reserve called the "Wabasca" area. At Victoria, drilling proceeded badly due to the problems which shale structures always cause. Only 750 feet were made in the summer of 1897.[21]

The natural gas found at Pelican River did not dissipate during the winter's lay-off. The well proved to be untameable by the drilling crew. For 1898, the crew managed to drill only 17 feet farther. The well was left to run its course. After running wild for twenty years, it was killed in 1918. The Victoria well was punched through without serious incident or return. When it reached the noteworthy depth of 1,650 feet, it was abandoned.[22]

The specific problem at the Pelican well and the more general failure to demonstrate Bell's theory about the pools of bitumen were major set-backs. The Survey's Director, Alfred Selwyn, admitted as much even as he clung to the logic of Bell's theory. Somewhat forlornly, Selwyn concluded that bitumen pools should be found through other experimental drilling, at least if a coherent explanation of the geology was valid. Selwyn did not face the possibility that a logical conclusion about the oil sands was not necessarily forthcoming.[23]

After 1898, no further drilling work was carried out. Perhaps the total bill of $38,000 (as much as a year's budget for the survey's field work) inhibited further exploration expenditure. The survey found many other projects and the Athabasca oil sands remained ignored. Clearly, however, no quick or cheap answer to the problems of developing the bituminous sands was likely.[24]

After the Pelican well was abandoned, government and private interest in the oil sands halted for about a decade. Between 1906 and 1908, some private drilling and promotional schemes were concocted. A number were initiated by a German immigrant, Alfred von Hammerstein, who took up drilling leases and worked the area near the present Bitumount site on the Athabasca River. His venture was marked by wild speculation, fraud and ultimate failure. It did lead to a period of intense activity which coincided with other pressures on the federal government and led to renewed federal investigations.[25]

Federal involvement recurred after 1912 in the context of British pressures on the Canadian government to devise a policy on oil exploration. In 1913, the federal government placed the oil sands region under a reserve, halting the speculative activity which had gone on in the previous half-dozen years. This policy decision occurred at a time when the Dominion was awakening to the importance of petroleum as a fuel. The British Admiralty was realizing the strategic and economic value of petroleum and began casting about for Imperial sources of crude oil. This British interest appears to have led Canada to generate a policy on petroleum development between 1912 and 1914. Alerted to the potential importance of oil and Canada's extreme dependence on foreign supplies, the federal government placed a federal reserve over two potentially productive regions, long identified by the Geological Survey. These were the Turner Valley, south-west of Calgary, and the Athabasca-Clearwater country, far to the north-east of Edmonton. Canada followed this by creating a development policy based on the strategic as well as the commercial role of petroleum. Private development was invited but explicitly subordinated to the government's right to expropriate lands and equipment. Moreover, only Canadian or British firms were allowed to take leases. Both of these conditions were abandoned after World War I. The notion that Mines Branch officials were awakened to the possibility of tar sands development solely because the resource was brought to their attention by Alberta MLA J.L. Coté in 1913 is mistaken, however; Coté's interest may be seen as a minor contribution to the decision to survey the tar sands that year.[26]

The survey was conducted by a Mines Branch geologist, Sidney C. Ells. Ells was son of R.W. Ells, a long-time Geological Survey official whose many studies included an examination of the oil shales of New Brunswick. This work had been carried out in 1902 when the younger Ells had acted as his father's assistant. After working as a coal mine and railroad surveyor and maintenance engineer throughout Canada, then graduating from McGill with a science degree in 1908, Ells joined the Mines Branch in 1912 as Secretary to the Director. He remained with the Mines Branch until his retirement in 1946.[27]

Ells's foray into the Athabasca country in 1913 was his first opportunity to conduct solo field-work and it provided him with the chance to prove his geological and engineering skills. Ells surveyed some 185 miles of river frontage and used a hand-auger to take two hundred core samples, investing a characteristically ferocious energy in his work. Ells argued that there was a resource of considerable extent which had great potential as a road-surfacing material; these asphalt reserves would constitute an immediately useful resource, he reported.[28]

Private drilling leases were taken up along the Athabasca River between 1906 and 1908; this primitive derrick was constructed about that time.
University of Alberta Archives

Following Ells's initial reconnaissance, he persuaded the Mines Branch to grant him funds to conduct an experiment in paving with the oil sands. But in an incident that would foreshadow his later status as a somewhat unreliable figure among oil sands experimenters and public servants, Ells ran afoul of his supervisors. He went ahead with his paving experiment in 1915 without heeding their instructions to defer to City of Edmonton engineers. Ells's official task was to procure oil sands supplies from the Horse River reserve, but he also high-handedly took over the supervision of the Edmonton paving operation. The paving operation was to be a joint federal-provincial-municipal venture, so Ells's failure to co-operate with personnel from the other jurisdictions did not leave the Mines Branch in happy standing with its partners. While the results of the paving operation apparently were satisfactory, Mines Branch Director, Dr. Eugene Haanel, strongly censured Ells for failing to account for some money spent and for ignoring the chain of authority. In his lame defence, Ells claimed the signs of a breakthrough in paving had led him to over-enthusiasm in conducting the work.[29]

Despite this setback, Ells managed to convince Haanel that he should continue oil sands work. In 1915, using samples he had dug and barged out from the McMurray area, Ells began further scientific examination of the oil sands at the Mellon Institute, a respected engineering school located at the centre of the American oil-and-coal-producing region in Pittsburgh, Pennsylvania. There, Ells worked with mining and petroleum engineers and chemists to examine the properties of the oil sands. In reporting the results of this work, Ells was to make clear a number of intriguing points about the oil sands and, typically, to create a furor that would have no mean impact on his later work and that of others in oil sands research.[30]

Ells typed his report, "Notes on Certain Aspects of the Proposed Commercial Development of the Deposits of Bituminous Sands in the Province of Alberta, Canada," early in 1917 before joining the Canadian Expeditionary Force. A huge, two-volume, unpaginated tome, Ells's "Notes" included twenty-three appendices, and reflected all the ungainly energy and enthusiasm associated with Ells himself. He surveyed the resource, its location and chemical make-up, and then examined its possible uses as a commercial road surfacing material and its refinement into a petroleum. While his report lacked any systematic characteristics, it did make a number of striking observations.[31]

First, Ells noted that the term "tar sands" was inaccurate. Quoting W.A. Hamar of the Mellon Institute, he argued that the sands were "bituminous sands," silicates impregnated with bitumen, the general term for all hydrocarbons, including tar. The bitumen was asphaltic rather than paraffin in composition. Thus, it was characterized as one of the two broad types of hydrocarbon compounds, deficient in hydrogen but profuse in carbon.[32] Ells confirmed the analysis of Christian Hoffman and the conclusion

Sidney Ells was a controversial figure in the era of oil sands research and development.

University of Alberta Archives

of Robert Bell that the oil sands indeed contained a normal type of petroleum product.

Ells went on to explain, in each volume, that the bituminous sands of the Athabasca were notable as by far the largest of the world's many deposits of bitumen-soaked shales and sands. This led him to discuss, at several points, efforts made in the United States, especially in California prior to 1900, to extract bitumen from deposits of sand. Ells could report only the signal commercial failure of such ventures, but he went on at great length trying to assess—somewhat inconclusively—the potential uses of the oil sands in road-building. He seemed to retreat from his earlier argument about the excellent prospects of the oil sands, although later he would change his views again.[33]

Ells's important conclusion was that methods of satisfactorily separating bitumen from sands existed; his corollary was that this extraction would be the key to any successful exploitation. Ells noted, using information acquired from his Institute associates, that three basic means could be used to conduct this separation of bitumen. One used petroleum or chemical solvents. A second employed heat distillation. The third utilized water.[34] Chemical and petroleum solvents as well as hot water had been applied in the California attempts Ells reported. In Appendix I of the "Notes," however, he gave pride of place to a description of "hot water separation" of bitumen. Ells claimed— although he presented no detailed data to support his argument—that bitumen 99.7 per cent pure was taken from samples washed with hot water and later subjected to a "filter press." While Ells admitted the need for further work, he indicated a strong sense of triumph in describing the results of the hot water separation work.[35]

Ells concluded his "Notes" with a plea for further investigation of the Athabasca deposits. He reminded readers that premature development could have deleterious effects on the area and its reputation. He also remarked that government ownership of natural resources and the 1913 decision to close the sands as a reserve were crucial policy decisions. There were, therefore, serious arguments in favour of continuing government research and development. He doubted whether private business would be able, let alone willing, to undertake highly experimental work under the limitations imposed by Crown control. Thus, only government-sponsored work could lead to the solution of the technical and economic difficulties which must be resolved before commercial use would occur. The goal must be commercial use, Ells insisted; the means must be government sponsorship.[36]

Ells's study occasioned further scrutiny of his work within the Mines Branch. At the behest of Director Eugene Haanel, branch consultants Karl A. Clark and Joseph Keele evaluated Ells's volumes. Their assessment was very critical. Clark, a chemical engineer, and Keele, a ceramics expert, saw the report as illogical and incoherent. Indeed, they claimed that their thirty-two-page review was necessarily couched in the form of a summary which tried to make sense of the report. They admitted Ells's research raised many interesting questions and that he had provided a useful summation of existing information. But they saw too many gaps in the data and too much incoherence in the report to permit publication, even if the material were rewritten. While Ells had noted some promising lines for attacking the potentially-great resource, he had not proven his case. Ells should rework his material, using sound scientific methods and reporting techniques. Theirs was a harsh critique, although it was not dismissive.[37]

The assessment was sufficiently scathing that Eugene Haanel acted to remove Ells from the project. Already annoyed by Ells's behaviour, Haanel reported to the deputy minister of the Interior, Robert McConnell, that Ells was incapable of the extra study that Clark and Keele demanded of him. Haanel stated that the geologist lacked the ability to conduct or write a sound scientific study.[38]

Just as Ells had been completing his "Notes" in 1917, Mines Branch engineer G. Parker commenced further study of the bituminous sands. Much more cautious and systematic than Ells, Parker confined himself to what he saw as the novel and highly promising potential use of the oil sands for road-building. Parker investigated this use because he saw it as technically feasible and economically possible. Parker conducted physical tests of the bituminous sand and compared the resource to commonly-used asphalts and road gravels. He found that similar "bituminous sand" deposits had been used for some years in

California. He noted that a graded mixture of tar sand and gravel had been an eminently utilitarian surface under great use for some years. But, if he was positive about the technical value of the road surfacing, he admitted that, even with Alberta's network of dirt roads, the economics of exploiting the resource would require much further work, including federal-provincial study.[39]

Parker's work, like Ells's, contained a warning to the Mines Branch that the oil sands required some extensive applied research. Thus advised, Mines Branch officials were cautious about entering into what was likely to be a tough undertaking. The dual impact of Ells's flawed work and Parker's careful investigation was such that the Mines Branch was hesitant about continuing research which would correct Ells's study or further Parker's. The federal department, then, shied from the project at the very moment when more impetuous Alberta scientists and public officials were eager to exploit the oil sands.

A less important, if equally notable, result of the Ells study was his estrangement from most of the major researchers in the area. The irony is that Ells continued to conduct studies for the Mines Branch of the mining properties of the oil sands and to assist in various paving operations in later years. But he never regained the opportunity to examine the major scientific problems of the oil sands or the engineering questions associated with separation, which he wished to do but for which he had no training. Little systematic documentary evidence exists to explain the impact of the Clark-Keele review and Haanel's stand, but there is little doubt that Ells was confirmed in his hostility by the decisions taken in the summer of 1917. Earlier, in spring of 1917, Ells had happily shared information about the oil sands with the University of Alberta chemist, Adolf Lehman. Once he returned to Canada after World War I and thence to his geological work in the oil sands, Ells was co-operative neither with the University nor with its offshoot, the Research Council of Alberta. After the Alberta research took shape, beginning in 1920, scientists in Edmonton thought that Ells was a man to avoid. The opinion of John Allan, who was lured to the university and then to the Research Council from the federal government, was typical: he did not want to work with Ells. This assessment was to be held by Allan's colléagues, as well as other business and government officials, for the many years Ells worked in the area. In 1924, Edgar

Stansfield, Honorary Secretary of the Research Council of Alberta, provided a summary of members' views about Ells to University of Alberta President Henry Marshall Tory. They were inclined to have nothing to do with him, a decision based solely on scientific grounds and not on personal bias. Perhaps Stansfield's distinction was ingenuous, but it did indicate the means which Alberta researchers took to neutralize Ells.[40]

Ells himself became convinced that various enemies in the scientific community were set on discrediting him unfairly. He seems never to have come to terms with the basis of opposition to his scientific work, let alone to his somewhat overpowering personal style. His reaction knew few boundaries of charity. He claimed to Alberta Premier Herbert Greenfield that Eugene Haanel had sought to discredit him because he thought he had unmasked Haanel's pro-German sympathies during World War I. He also wrote to Greenfield that Karl Clark, who had joined the Research Council of Alberta in 1920 to work on the oil sands, was motivated solely by the desire to oust Ells from research in an area where he had pointed the way. Ells hinted that Clark's work was derived from his own, which thus explained Clark's desire to isolate him. Ells continued to feel and express his sense of persecution and his intense rivalry with Clark during the remaining thirty years of his active life. His numerous publications on the bituminous sands contained virtually no references to Clark's role even when he noted the Alberta work.[41]

While discussion of reaction to Sidney Ells's work and his relations with his research colleagues may seem trivial and amusing, it is not insignificant. If only to clear up confusion about Ells's role in the oil sands development process, it is useful to know just what he did up to 1917 and how it was received. It is also useful to explain why the federal initiative, active from 1913 to 1917, became hesitant after the war ended. The considerable uproar surrounding Ells must be seen as an influential factor in the shift in initiative from federal to provincial governments and the relations between research groups in the 1920s.

* * *

The works of the Geological Survey and the Mines Branch were important in three ways.

First, Geological Survey studies were essential in establishing the characteristics and possible significance of the bituminous sands. If Robert Bell's theory about pools of bitumen was eventually proven false, his point and Hoffmann's analysis about the bitumen content of the sands and its possible uses were most important. Perhaps because of the fact that such impressive authorities as Bell and McConnell, working for an equally credible agency of government, proclaimed the importance of the resource, it was accepted as a subject for research and a source of petroleum. Had no such authoritative men and institutions been involved, it seems doubtful that the later work could have been undertaken as readily as it was.

Second, in a more negative way, Bell's and McConnell's work stimulated all of those drillers, from Hammerstein to Fitzsimmons, who vainly sought an elephant pool of oil in the Athabasca country. Their labour was a testament to the importance of the Geological Survey's work. Similarly, the remarkable $38,000 cost of the drilling on the Athabasca and Saskatchewan Rivers was an ominous portent of the way in which experimental work was so often to become unsupportable. This raises the point that the Survey, and later the Mines Branch staff, recognised and had warned of the problems and costs which resulted from work in an isolated area and on an intractable resource.

Finally, and most positively, Ells's attempt to persuade the Mines Branch to redirect its work on the oil sands was a catalytic factor. However imperfectly, Ells summarized knowledge about the oil sands, made clear that bitumen must be separated from sand, and indicated the many gaps in knowledge of geology and engineering which existed. His flawed work pricked the attention of others, both private developers and research-oriented scientists like Karl Clark and Henry M. Tory. They were to press on with further work in the 1920s, work which would clear up some of the problems Ells had noted, as well as some he had created.

30

II: A Provincial Initiative: The Research Council of Alberta 1919-1930

If the reaction to Ells's sprawling report was unfavourable, it was not dismissive. Ells's study led other engineers and geologists to conclude that more oil sands research was necessary. The bituminous sands certainly constituted a valuable source for bitumen and the bitumen might well have petroleum as well as asphaltic applications.[1] Particularly interested in these findings were some ambitious academics and politicians in Edmonton.

A government-supported agency, the Research Council of Alberta, was formed after World War I to assist the province in pursuing broad goals of economic development and diversification. Among the industrial and scientific research studies undertaken was one of the oil sands. Karl Clark moved to Edmonton to conduct his work. During the twenties he engaged in the task of separating bitumen from the oil sands, and he attracted talented assistants to work with him, among them Sidney Blair and David Pasternack. Clark's work led to the construction of a bench-scale separation plant at the University of Alberta laboratory, then a larger one-man plant built in north Edmonton and, finally, the completion of a field-scale separation plant at Fort McMurray.

* * *

The Scientific and Industrial Research Council of Alberta was formed in 1921. It resulted from the efforts of academics and politicians interested in creating an organization which would conduct studies on resources and processes which might permit diversification of the province's economy, thus freeing it from its dependence on agriculture. A series of meetings took place at the University of Alberta in 1919 and these led to the creation of a committee which designed the new research agency. The committee was comprised of the university's president, Henry Marshall Tory, a scientist himself; N.C. Pitcher, professor of engineering; John A. Allan, professor of geology; John Sterling, provincial inspector of mines; and Provincial Secretary J.L.Coté, an enthusiastic advocate of exploring the Athabasca region. Together, these men defined the Research Council's general purposes and its areas for research, which included coal resources, mineral deposits, and the tar sands. The Research Council was the first government research council in Canada, although it closely resembled a federal government research "committee" created during World War I, which became the National Research Council in 1928.[2]

While the Research Council was in its formulative stages, a joint federal-provincial proposal to co-ordinate research into Alberta's coal and tar sands resources was advanced in 1919. (The federal government continued to administer the natural resources of Alberta, Saskatchewan and Manitoba, the provinces created out of the former Northwest Territories until 1930.) Despite the logic of a co-operative federal-provincial approach to research into natural resources, there was contention between the parties involved. In the spring of 1920, President Tory, angry about federal officials' estimates of the value of Alberta's tar sands and other resources, disengaged the province from formal co-operative research projects. Tory's reaction was typical of that of other University of Alberta and Research Council officials towards federal bureaucrats, and what was perceived to be their failure to place adequate importance on the investigation and development of Alberta's resources.[3]

Tory thought, as he later put it for Alberta Premier Herbert Greenfield, that state-funded scientific research would be "laying the foundations of accurate knowledge upon which we can build our industries with security in the future." To Tory, secure private development depended on prior direction by the publicly-funded Research Council.[4]

Even before the Research Council was formally created, President Tory received Provincial Secretary Coté's approval to conduct research into the tar sands. Accordingly, in June 1920, Karl Clark was appointed to study the extraction of bitumen from the oil sands and to determine the value of bitumen as a road surfacing material.

Clark was well-qualified for the work. After graduating from McMaster University in science in 1912, Clark went to the University of Illinois, where he completed an advanced degree in physical chemistry in 1915. He then worked at the Road Materials Division of the Mines Branch in Ottawa. Clark remained at the Research Council from 1920 until it suspended operations in 1935. After teaching part-time and working in industry, he returned to Edmonton to teach mining engineering at the University of Alberta until he retired in 1954.[5]

Clark was following in the footsteps of other University of Alberta scientists when he set about his investigations of the tar sands. In December 1912, W.M. Edwards, a professor of engineering, had encouraged Tory to permit faculty members to acquire tar sand samples from the Mines Branch. Some experimental paving using untreated tar sands appears to have taken place at the university after this, and in 1917, Adolph Lehman, a professor of chemistry, told Sidney Ells that he had been studying the sands' basic chemistry. Lehman's studies provided Tory with essential information as he tried to expand research.[6]

Lehman also supported Tory's angry reply to a review of the potential of the tar sands sponsored by the federal government's industrial research committee in 1920, at a crucial juncture in federal-provincial negotiations over joint research. Tory was informed by the federal committee's A.B. Macallum that two McGill University scientists had found that the tar sands contained "ingredients belonging to the naphthalene series," bitumen useful only for producing asphalts and not for "petroleum oils." In response, Tory drew on Lehman's work to point out that bitumen from the tar sands had the properties of asphaltic hydrocarbons, such as California petroleums, although not those of the paraffin hydrocarbons found in Pennsylvania, which would have been more familiar to the McGill chemists. Armed with this evidence (which accorded with the common chemical properties of all bitumens, unlike the dubious

Karl Clark's involvement in Alberta oil sands research and development commenced in 1920 when he was appointed to study the extraction of bitumen and to determine its value as a road-surfacing material. Clark, a much respected figure in his field, remained active in it until his retirement in 1954, sixteen years after this photograph was taken.

Provincial Archives of Alberta

distinctions made in the McGill study), Tory scoffed at the report presented by Macallum and reiterated the Alberta decision to proceed with research into the oil sands.[7]

At the time when the Research Council engaged Clark, international concern about the dependability of petroleum supplies was intense. Immediately following World War I, petroleum engineers and economists, with all the unanimity of their expertise, had concluded that the rising demand for crude oil would lead to serious and chronic shortages of supply for the foreseeable future. The prospect of world oil shortages at a time of rapidly growing demand informed the prevailing wisdom of oilmen everywhere. Their expectations would remain industry opinion until the mid-1920s when, all too typically for the oil industry, shortages turned to surpluses and prices plummeted. The discovery of new oilfields in California and Oklahoma drove North American prices downwards, while the realization that Middle East oilfields constituted an immense resource led the international oil companies to create an effective programme of price-fixing and supply management. By the late twenties, the problems of falling prices and glutted supplies dominated the industry. From an international perspective, the tar sands were a highly alluring resource in 1920, an interesting one by 1926, and irrelevant amidst the price collapse of the early 1930s.[8]

If international shortages seemed imminent in the post-World War I era, Canada's dependency on international supply was almost total in the twenties. There was only minor oil production in Canada, with tiny amounts from Turner Valley and New Brunswick, while the largest portion came from the old Ontario fields near Sarnia. A federal government bounty had been placed on domestic production after 1890 to encourage output, but the simple fact was that there were no known large oil reserves in Canada in the early years of this century. As a result, Canadian production filled less than 1 percent of Canadian needs in 1921. Even after the Turner Valley oil discoveries of 1924 and 1925 led to a dramatic increase in Canadian output, Canadian oilfields supplied only about 4 percent of domestic requirements in the late 1920s. (See Table 2.1.) Thus, even when Turner Valley output was more than one million barrels a year, as it was by 1931, total Canadian consumption that year was more than 35 million barrels. (See Table 2.2.) The post-1925 rise in Turner Valley production meant that the tar sands became somewhat less significant insofar as local markets were concerned. Turner Valley producers had a difficult time competing with the cheap foreign oil which was increasingly available. The possibility of the tar sands providing a large petroleum source for Canada remained, but the reality of the market shifted against the resource during the 1920s.[9]

Karl Clark, however, began his research with deep hope at a time of concern over dwindling supplies, and with no mean knowledge of his subject, for he had read Sidney Ells's compilation of the work done at the Mellon Institute. Clark worked quickly and soon he reported that he understood how to separate bitumen from the oil sand.

Table 2.1 Crude Oil Supply, Canada Domestic Production, 1911-41 (barrels)

Year	Domestic Production bbl.	Imports bbl.	Total Supply bbl.	Domestic Total %
1911	291 092	3 339 800	3 630 892	8.0
1916	198 123	8 355 029	8 553 152	2.3
1921	187 541	19 714 286	19 901 827	.9
1926	364 444	22 153 286	22 517 730	1.6
1931	1 542 573	35 358 029	36 900 602	4.2
1936	1 500 374	35 835 264	37 335 638	4.0
1941	10 133 838	146 791 026	156 924 864	17.8

Source: Canada, Department of Mines, "Mineral Production of Canada"; Canada, Dominion Bureau of Statistics, "Crude Petroleum and Natural Gas Industry in Canada."

Table 2.2: Alberta and Canadian Petroleum Production, 1917-35 (barrels)

Year	Alberta	Canada	Alberta/ Canada %
1917	8 500 (est.)	206 899	4
1918	13 040	304 741	4
1919	16 437	240 466	7
1920	11 032	196 251	6
1921	no report	187 541	not applicable
1922	6 559	179 068	4
1923	1 943	170 109	1
1924	844	160 773	0.5
1925	183 491	332 001	55
1926	216 050	364 444	59
1927	318 741	476 591	67
1928	482 047	624 184	77
1929	988 675	1 117 368	88
1930	1 398 160	1 522 220	92
1931	1 413 631	1 542 573	91
1932	906 751	1 044 412	87
1933	995 832	1 145 333	87
1934	1 253 966	1 410 895	89
1935	1 263 510	1 446 620	85

Source: Canada, Department of Mines, "Mineral Production of Canada"; Canada, Dominion Bureau of Statistics, "Crude Petroleum and Natural Gas Industry in Canada."

As early as August 1921, Clark wrote to Tory that "something definite has been accomplished and a very considerable glimmer of daylight let through the problem." By the autumn of that year, Clark saw his problem as one of determining how best to demonstrate and publicize his work, rather than how to improve his methods.[10] No wonder Tory was so hopeful himself.

Clark's early optimism was remarkable and must have struck him from time to time in the long years in which he worked at the problem of extraction. But in 1921 he announced that

> most of the purely inventive work has now been done. There remains to be accomplished the practical application of the new method to the production of bitumen from the tar sands. This means .... on a commercial scale.

He used an arch and even secretive language in explaining this "new method," but he certainly felt that his breakthrough was undoubted. Clark had aimed at the "practical exploitation" of the resource and found the only extraction method which would work. He argued—a bit unfairly to other researchers like Sidney Ells—that he had found a novel method distinct from the two well-known ones of "retorting" bitumen by heat and using a solvent-distillation technique. Clark pointed to his successful development of a hot water separation process as being his own. Clark's expenses for his initial work in 1920-21 amounted to $300 for laboratory equipment. [11]

Clark's position in 1921 is worth pondering for two reasons. First, he made an exaggerated claim of discovery. As we have seen, the much-maligned Ells had worked on this concept with the help of Mellon Institute scientists and it was then but one adaptation of previous work on the problem. Moreover, hot water separation was a common mining technique, which is why it had been used in the California oil shales work of the 1890s. Clark was ignoring the cumulative nature of his technique, for reasons that seem unlikely to yield documented explanation.

The second reason is that Clark was excessively hopeful. He saw the next stage of work as the tidy completion of the project. As he noted, a large-scale, field-sized, separation plant and further studies of the uses of bitumen (i.e. petroleum and road materials applications) were all that were required. In the event, many years were to intrude before Clark's zesty goals were achieved.

Clark persisted in his laboratory work in 1922, hesitating to publish his findings due to concern about the impact it would have on actual development and about the means to gain priority for patenting his work.[12] Perhaps, too, he was becoming a bit hesitant to press his earlier optimistic and aggrandized claims. But he led Henry Marshall Tory to tell Charles Stewart,

The Research Council of Alberta Tar Sands Separation Plant, University of Alberta power plant basement, Edmonton, 1923.

University of Alberta Archives

Liberal MP and Alberta's cabinet minister as Interior Minister, that work was well advanced. "We have solved the problem of the separation of bitumen," Tory wrote,

> *in so far as its relation to the problem of roads is concerned. Clark has succeeded in getting out by a very simple process nine tenths of the sand in one operation.*[13]

This summation of the laboratory separation was true enough, perhaps, but Tory might have lingered over the words "solved" and "simple process," which did not describe the sorts of problems in perfecting the basic process that would take thirty more years of research.

Clark's printed reports for the Research Council's annual volumes were far more cautious than his private writing. In 1922, he wrote that the actual bitumen source was *unlikely* to be amenable to petroleum cracking. But in both 1922 and 1923, he emphasized the value for asphalt products. Thus in 1922, Clark elaborated on the chemical distinction of "bituminous sand" bitumen from paraffin-based petroleums and from other asphalt-based petroleums like Trinidad asphalt and American oil shales. He then explained why solvent or distillation methods of separation were not appropriate. He concluded that another method was needed. In 1922 he was silent on this alternate method, although anyone familiar with the problem might have guessed that he meant a hot water treatment process. In 1923, Clark announced the development of a hot-water-washing and silicate-of-soda mixing process as the key to separating bitumen from oil sands. However, he must have been chagrined to admit that an English chemist, Ernest Fyleman, had published recently and filed patent on the method Clark had thought was his own. Clark could only go on to claim—rightly—the need for much engineering and chemical research to find the best mixtures of oil sand and silicate-water solution and best method of pulping and separation to ensure that the process was most efficient.[14]

Only in 1927 did the Research Council begin to print the long-awaited scientific reports on the "breakthrough" that Karl Clark since 1921 had seen as his vital contribution to research. The publications were Clark's first extended scientific works. The Research Council's Report No. 18 on "The Bituminous Sands of Alberta" was printed in three small volumes, summarizing Clark's work from 1923 to 1925. The first two books were written by Clark and his assistant, Sidney M. Blair; the third, printed in 1929, was written by Clark alone. The first volume, on the "Occurrence" of the sands, recorded the results of the Clark and Blair survey of the geology and chemistry of oil sands, and contained a description of the development of exploration and current methods used to analyze the sands, complete with maps and tables.[15] It is a readable and informative survey of the tar sands and their early reconnaissance. Their goal was nothing less than "the commercial development of the bituminous sands." This followed all Research Council thinking since the institution's inception. Clark and Blair listed the numerous promotions of private developers since 1920 with the warning that no breakthrough in commercial application had occurred. They did claim that more persistent scientific work had finally

A shipment of raw bituminous sands is unloaded at Karl Clark's plant in the Dunvegan Railway Yards in 1924.   Provincial Archives of Alberta

yielded the "basic information" that "may soon" lead to industrial activity and commercial success.[16]

Long summaries of the Research Council tests on various site samples (Blair had conducted the field work which garnered the core samples) led the authors to note that the bitumen content of oil sands was almost always between 10.5 percent and 13.5 percent of total volume. Moreover, both extensive silt and clay content of surface deposits and a 4-6 percent sulphur content would disturb any extractions.[17] Still, the key points that bitumen was recoverable from very large and rich reserves and that a full range of petroleums was refineable from the bitumen portended well for commercial development.[18]

In the second volume, "Separation," Clark and Blair discussed the development of Clark's version of hot water separation (or extraction, as engineers call it today; see Appendix). They explained that a separation process was crucial in creating a marketable product for either road materials or petroleum products. They mentioned the "outstanding" road material base that untreated oil sand could provide, but slipped over any but anecdotal proof of this contention. As they admitted, the separated bitumen was "more readily" adaptable to road surfacing uses than the raw mixture of bitumen and sand. The authors moved quickly, then, to examine the bitumen as a product suitable for "cracking" into petroleums. To extract the bitumen for further refining, the hot water method, after all, was best suited.[19]

It is striking to contemporary readers how fixed Clark and Blair were upon the commercial potential of their research. As in Clark's private musings, his published work defined the purpose of research as being to find the quickest method that would commercialize the product. In the preliminaries to the second Clark-Blair volume, estimates of freight charges to ship bitumen were compared with the costs for conventional asphalt. This led them to indicate—but not to assert—that freight comprised the key limitation on the tar sands and its products.[20]

In any case, the study of separation explained the basic and improved technique of hot water separation. The authors admitted the conventional quality of the process in treating other, smaller tar sands deposits. They did not mention that hot water washing was also an adaptation of standard mineral-refining techniques. They explained the key to their particular process, the one which had so elated Clark in 1921:

> When bituminous sand which had been kept in contact for some time with a hot dilute solution of silicate of soda was introduced with agitation into a comparatively large body of hot water, a complete dispersion of the bituminous sand took place....

Karl Clark and Sidney Blair were responsible for the design of the Research Council of Alberta Tar Sands Separation Plant at Dunvegan

Clark and Blair argued that the process worked because the "water-in-oil emulsion" of silicate of soda and bitumen could be flooded with hot water to produce an inversion. Thus an "oil-in-water" emulsion resulted. In that case, "minute globules of bitumen form and rise to the surface of the water," and these could be skimmed off as a relatively pure bitumen.[21]

The study included a summary of the operation of two plants which Clark had built, along with their results. In 1923, a small separation unit based on Clark's design was built in the basement of the University's power plant. The machinery processed some eighty-five tons of sand. The results were encouraging in that water-free bitumen with 8-12 percent of mineral matter was produced. Yet problems with continuous operation and very high heat costs limited this initial work.[22]

Clark designed a larger-scale plant which was constructed in 1924 at the Dunvegan railway yards on Edmonton's northern limits. This plant was more sophisticated, being capable of simultaneous washing of tar sand and a separation of the bitumen from the water-oil sand mixture (see Appendix: Historic Plant Developments). While the plant operated poorly because there were numerous design flaws, it was possible to process about 100 tons of sand. The "refined" bitumen product contained a large mineral content, some 30 percent on average, compared with 9 percent for the basement operation of 1923.[23]

Yards, Edmonton.   Provincial Archives of Alberta

Clark and Blair engaged in considerable winter labour during 1924-25 to overcome the problems of the first Dunvegan plant. The two engineers worked on changes to the initial treatment process, the addition to tar sands of the silicate of soda, so that better mixing would occur. Moreover, the authors warned that the use of "weathered" sand in 1924 had reduced the likelihood of success. Only "fresh" sand was readily refined and only fresh sand would be used in 1925.[24]

The Separation Plant at the Dunvegan Yards in 1925.

Provincial Archives of Alberta

With this information and some changes, the 1925 season saw the operation of a much rebuilt plant at Dunvegan. The key change of using "a second-hand clay mixing machine" adapted with steam coils and water pipes meant that primary treatment was far more efficient. The plant ran smoothly, Clark and Blair claimed. The separation yielded a bitumen with but 7 percent mineral content on average. About 500 tons of sand were processed during the season from a plant which contained more complex and more numerous parts.[25] (See Appendix: Historic Plant Developments.)

It was at this stage that the researchers endeavoured to estimate the cost of their separation process, a brave attempt given the limited scale of the operation, but one characteristic of Clark and Blair. Proving the commercial value of their efforts and the product was their foremost goal. Clark undoubtedly used the results of his 1926 tour of various tar sand and shale oil sites to piece together his estimates about the necessary scale of operations. The resulting estimate of cost was based on consideration of mining costs for a steam shovel working 200 10-hour days per year and separation costs for a 1,000 ton per day plant. This sophisticated doodling led Clark to show that an annual production of 27,000 tons of bitumen (based on processing over 200,000 tons of sand) would cost about $1.00 a barrel. The plant was to cost $200,000 to build. But, Clark included no transportation charges, let alone financing costs in his estimates. Still, Clark and Blair claimed that the pilot plants and the extrapolated 1,000 ton per day plant showed that separation was effective, efficient and economic.[26] The fact that this projected plant would process twice as much sand in one day as they had in several weeks indicates the scope of their extrapolation.

The rest of the Clark-Blair report and the later volume written by Clark alone contained additional discussions of the uses of bitumen. In his discussions, Clark argued that motor fuels and asphalts and road oils were the useful by-products that would lead to commercial development. By then aware of the assessments of the industry at Universal Oil Products in Chicago and Canadian Oils Limited in Sarnia, Clark himself concluded that the road material uses for bitumen were more limited than the possible motor fuel use. He did so because he saw the economic limitations of development as the crucial ones. As he stated, "technical limitations do not stand in the way" of development. Still, Clark was vague about the technical advantages of bitumen for either road asphalt or motor fuel, although more assertive about positive motor fuel analyses. It can be concluded that he was well aware how much further work was needed to prove the value of the resource, despite the advance of his primary separation work.[27]

Following his 1924 and 1925 work with Blair, Clark engaged in little substantial tar sand separation experimentation in the three following years. Sidney Blair, in fact, left the Research Council for Universal Oil Products in Chicago in 1926. He had earned a Master's degree in mining engineering at the University of Alberta while working for Clark, and he sought opportunities for increasing his income which were not then available in Canada. His international career flourished for the next twenty years. Clark was left to conduct desultory paving and other analytical operations as well as to complete reporting for publication the results of the research he had done with Blair.[28]

In 1926, Karl Clark took a two-month grand tour of American and Canadian refineries as well as of highway construction operations in the United States. This trip was notable for two reasons. The first was that Clark established links with refiners such as the Chicago petroleum refining firm, Universal Oil Products. This company was engaged to examine the "cracking" characteristics (the refinable petroleum products) of bitumen produced from tar sand. Clark also sought out Canadian opinion. Canadian Oils Limited of Sarnia also tested Clark's bitumen samples. These tests were conducted by a company employee, Earl Smith, who was later associated with the Abasand project; Smith reported that the bitumen was much like a "heavy" Texas crude, excellent as a lubricant but not good as gasoline. The American firm's research facilities and studies apparently were more complete, for those were the results most valued by Clark. They sustained Adolf Lehman's views as well as Clark's about the totally acceptable quality of bitumen refined into motor fuels. The "heavy" oil was certainly useable in a number of ways.[29]

The second important result of Clark's trip was that he discovered how uninterested oil refineries were in the actual production of refined oils from his bitumen resource. The story was the same in Chicago and Toronto. Increasing oil supplies, due to more efficient drilling methods as well as discoveries, and falling oil prices meant that there was no interest in the tar sands as a source of petroleum. The oilmen encouraged Clark to continue his research: the product was a good one and long-term market and supply factors suggested that the tar sands would be useful. But not yet.[30]

It is little wonder that Clark devoted the bulk of his tour to the current state of road construction and asphalt applications in the United States. He was unshaken in his view that the Research Council's studies had been thoroughly correct and timely in the question of road surfacing. That at least showed the value of the previous years' work.[31] It should be no surprise, however, that

Clark's findings on his tour resulted in a moratorium on tar sands work.

By the time Clark had prepared a report on the results of the work he and Blair had conducted for publication in 1927, his old optimism was rising again. By the end of 1927, he was convinced and outspoken about the direction his work should take, and the reasons.

Meanwhile, Clark had conducted some experiments on the paving uses of bituminous sand and separated bitumen and had observed several others. Clark had witnessed several projects completed before 1923 with some skepticism. While he did not reject the quality of the work, neither did he endorse it. Moreover, he produced information based on City of Edmonton investigations in 1922 which showed how uneconomical bituminous sand sidewalks were compared with ordinary asphaltic sidewalks. The difference was just under 2 cents per square foot, 17.4 cents to 15.6 cents. Clark concluded that sidewalk paving was impractical—enraging entrepreneurs like Tom Draper who were promoting this work.[32]

Clark was more hopeful about the road surfacing uses of the bituminous sands. But even in his 1929 summary of his road materials research, he was guarded in assessing the technical quality of bitumen as a basis for asphaltic road surfacing. He emphasized that the extent of prairie road building would prevent sufficiently widespread use of bitumen in road surfacing. However, he was at least tacitly reiterating the results of his own experiment of 1927 on a six hundred-foot stretch of the Edmonton-St. Albert "Trail" and his observations of other work. These results had shown Clark that only separated bitumen would have any serious use as a road surfacing material. His attempt to use separated bitumen had not been very successful due to its excessive viscosity; this inhibited its mixing with gravel. Clark had experimented in creating an emulsion of bitumen and water to enhance the mixing of bitumen and gravels as an oiled road surface. This emulsion had not resulted in an effective bonding between the bitumen and the gravel, however, and the consequence was that the surface broke down into its constituent parts over a period of use. Nonetheless, Clark remained convinced that further work in establishing a better bonding might well succeed. His positive assessment of 1929 remained something of a hollow claim in light of his work in 1927.[33]

Karl Clark conducted experiments into the paving uses of bituminous sand and separated bitumen. Shown here is a "bituminized" earth road, part of the St. Albert Trail, in 1925. Provincial Archives of Alberta

Sidney Ells, too, had conducted further work on paving during the twenties. He supervised the "surfacing"—the application of a heated mix of raw bituminous sand with gravel—of some three miles of roadway in Jasper National Park. This road required frequent re-rolling, for it was not a hard surface, although Ells claimed it was similar to a macadamized road. Ells's conclusion was that the roads surfaced in this manner were serviceable, better than the dirt roads which predominated at that time, and that this paving material was cheaper to lay down than conventional asphalt and gravel (93 cents per square yard compared to $1.25).[34]

Karl Clark, not surprisingly, denied the cheaper procurement costs and doubted the technical standards of any raw mixtures of oil sands. He only allowed that the separated bitumen should meet technical standards as an asphalt paving base. But even that depended upon the refining which normal asphalts underwent before use as a road surfacing base.[35] In this way, Clark again pointed to the necessity of developing industrial demand which would lead to the refining of the crude bitumen. And that necessity, in turn, led to the goal he had decided was essential, the use of bitumen as a base for motor fuels and lubricants.

In late 1927, Clark initiated a new research offensive against the oil sands. His elaborate reports sustained his goal of finding the basic process by which bitumen could be separated from its sand mixture. His private statements were even more categorical. "A process has been worked out," he wrote to the Research Council board of directors in 1927 (just before his final report on "utilization" was finished),

> *and has been demonstrated by fairly large scale operations to be practical. All data so far obtained indicate further that the process is economical. The basic general problem of bituminous sand separation has been solved.*

Clark, then, asserted that research had reached the stage where technical considerations were minimal and economic ones crucial to further development. The fact that no serious exponent of his method or goals yet had emerged must have also preyed on him. To demonstrate further the practicality and feasibility of development of the oil sands, Clark proposed that the Research Council support a new round of full-scale study. This time, he sought the construction of a good-sized plant to be built on the site of oil sands deposits.[36]

Clark's new research direction was complex. He had to obtain permission to hire a chemist as his new assistant, a belated replacement for Blair. This was achieved when David S. Pasternack was appointed in 1928. Pasternack was a Latvian-born graduate of Queen's University and of McGill, where he received a Ph.D. in Chemistry in 1928. He worked for the Research Council and in private business, retiring from the Council in 1969. Clark listed a series of questions about the use of further oil sands supplies, still held by the federal government, of course. Finally, he estimated the cost of building a field plant, which would be located about three hundred miles north of Edmonton. Legal and financial considerations, he realized, meant that Alberta and Ottawa would be forced to co-operate in any further projects of his in some way or another.

In early 1928, Henry Marshall Tory moved from Edmonton to Ottawa to preside over the newly-organized National Research Council. A geologist from the University of Manitoba, the Scottish-born and educated Robert C. Wallace was Tory's replacement as president of the University of Alberta and director of the provincial Research Council. Alberta was still eager to continue developing its northern research and Wallace's appointment enhanced this drive, for he was a respected expert in this area, having been chairman of the Northern Manitoba Commission in the twenties. As well, the long negotiations between Alberta, Saskatchewan, Manitoba and the Dominion government were about to yield the transfer of natural resource jurisdiction to the provinces. The province, led by its able and committed Premier, John E. Brownlee, who was also chairman of the Research Council, supported in 1928 a Natural Resources Act to finance northern research and to prepare for provincial takeover of natural resources.[37]

Given this convergence of circumstances and sympathetic individuals, conditions now seemed right for more effective co-operation between provincial and federal officials and politicians. In May 1929, a two-year co-operative agreement was signed between the Research Council of Alberta and the Mines Department of Canada. It seems significant that Robert Wallace of the Research Council and Charles Camsell, then deputy minister of Mines, were joined as administrators of the project by

David Pasternack (centre) became Karl Clark's assistant in 1928.
University of Alberta Archives

Henry Marshall Tory, the National Research Council's representative.[38]

The "Bituminous Sands Advisory Committee" was to finance
and administer two years of research. Alberta was to work on the
construction and operation of a separation plant at a suitable
site in the Athabasca region; in addition, there was to be
consideration of the petroleum refining uses of the product. The
federal group was to carry out research on the raw material, to
mine, and to supply samples from the federal government
reserve for any Alberta plant; it would also support a road-paving
programme. This joint venture reunited Clark and the Research
Council with the Mines Branch, including Sidney Ells, in one
grand effort to demonstrate all the technical knowledge about
the value of the oil sands. It might be noted that this joint
venture was one of several in which effective federal-provincial
co-operation occurred in the late twenties. Other areas of work
included voluminous studies of coal resources and an
examination of possible uses for Turner Valley natural gas (the
flaring off of up to sixty million cubic feet per day constituted one
of the most wasteful uses of a natural resource in the nation's
history). It appears that the move toward provincial control of
natural resources—and the effects of a decade of prairie political
protest through the Progressive party—led to a co-operation
which was only sensible, yet all too rare.[39]

The Research Council of Alberta had settled into an organized
and systematic role by this time. Its main research divisions—
Fuels (coal), Geology, Soils, Natural Gas, and the misnamed
Road Materials (tar sands)—were engaged in a broad range of
basic studies of the province's natural resources and their uses,
or waste in the case of gas reserves. Without making an
assessment of the scientific or direct economic value of this work
(serious study had yet to be done on these questions), it seems
fair to conclude that the Council was active and was deemed so
by government and scientific groups, as well as by business. The
appropriations voted for the Research Council were never large,
however. It received $36,000, including $20,000 for expenses and
$16,000 for salaries in 1923 from the Alberta budget of $9 million.
By 1930, the Council received over $85,000, including $45,000 for
operations and the rest in salaries. The Alberta budget that year
was $15 million. But the Council's importance was reflected in

the facts that the Premier himself was chairman and that the board included both cabinet members and the University President.[40]

The Council was reorganized in late 1929. It had the precise and hopeful goal of advising the government on methods to expand industry and develop resources. This sustained the areas of work developed since 1921 and reflected the expectation of imminent transfer of resource ownership and a decade of generally favourable economic conditions. Like the oil sands work, the larger scientific and industrial enterprise sponsored by Alberta was renewed just as the world's economy was edging toward a prolonged slump, much to the province's considerable impoverishment for over a decade. Indeed, one result of this long-term and world-wide development was the suspension of the Research Council itself in 1933; it was not revived for a decade.[41]

Redoubled efforts in oil sands research in 1929 involved first, redesigning the Dunvegan separation plant and, second, dismantling and rebuilding it on the Clearwater River near the Fort McMurray suburb of Waterways, the railhead of the Northern Alberta Railway. The plant was not ready for operation until 19 October 1929. The reconditioned plant had a 60 tons per day capacity. Only 11 barrels of bitumen were produced in the 3 days of operation in the autumn of 1929. The bitumen was of high quality, containing less than 10 percent mineral content, but over 30 percent water. Karl Clark admitted that an emulsifying agent used to make separation effective was not working, but he felt pleased that field operations had been conducted. He later noted that his chemist-assistant, D.S. Pasternack, who was in the University of Alberta laboratories, had begun to perfect an emulsifying and separating process. Pasternack's tests had produced bitumen with only 2 percent mineral content and 6 percent water content, both excellent results. This contrasted with the usual 10 percent minerals and over 20 percent water in previous bench-scale tests. Clark was beginning to surrender to highly optimistic assessments about the capacities of the Clearwater plant.[42]

After a winter of study, Clark, Pasternack and their millwright, John Sutherland, began working on the Waterways plant in May 1930. The plant's initial runs were a great disappointment, as

The Research Council of Alberta Clearwater River Plant under construction in 1929. University of Alberta Archives

even the tar sands feed hopper failed to operate smoothly. When the plant ran for several hours at a time, as it did on 7 June, when it processed 20 tons of sand in eight and one-half hours, there was a terrible waste of steam in its various operations. This slowed the whole process. Equipment failures and inefficiency plagued the operators throughout the summer. Still, on 24 June, Clark noted that the plant was at last working properly. It had produced that day, he wrote, a crude product "rich in bitumen ... soft and nice. The tar rolled out of the plant in great style...." Then the wheel which removed the bitumen froth stopped picking up froth. Many days and many adjustments later, Clark fastened on two key problems. A continuous and consistent supply of tar sand required the installation of a "pug-mill" (a machine to blend materials into uniform consistency). Mixing a more effective emulsifier intrigued him; silicate of soda was still used, but the exact proportions seemed to require constant adjustment. The plant also continued to consume steam too wastefully and to operate quite unreliably. But it did produce bitumen. Some 800 tons of oil sands were processed from June to late August 1930, yielding some 75 tons of bitumen, over 15,000 gallons. Most important, Clark later wrote, the quality of bitumen was very good, with just 5 percent mineral content and—no thanks to an unreliable dehydrator—only 1 percent water content.[43] Aside from problems with salt and rock content in the raw sands (ironstone modules often jammed the feeder), the plant on the Clearwater had poured out good samples of crude bitumen.

The plant had worked but it had bedevilled Clark and his

associates throughout the summer of 1930. Clark confessed in his log that he was frustrated by the plant's unreliability. The product, however, was most encouraging, comparable to Pasternack's laboratory results, thus proving Clark's claims to

The Clearwater River Plant processed some 800 tons of oil sands between June and August 1930; these oil sands yielded 75 tons of bitumen.
Provincial Archives of Alberta

have solved the basic technical problems. These small amounts of bitumen tested out much like the products of the last Abasand and the successful Bitumount plants built at much greater cost in the forties. In its way, the field station was a success. Karl Clark concluded in the Research Council's Annual Report of 1930 that

> *a separation plant has been built and successfully operated in the north country at the deposits. This has had psychological as well as technical values . . .*[44]

Evidence of the psychological breakthrough appears in the interest shown by two American oil engineers, J.M. McClave and Max W. Ball, who were convinced to take up the challenge of developing the oil sands. The tar sands tinkerer, Robert Fitzsimmons, had also been drawn to rework his approach to the tar sands. A Toronto developer, W.P. Hinton, also expressed interest in buying the Research Council plant and acquiring oil sands leases.

Given hindsight about the economic depression then beginning to settle upon Canada, it is surprising that there were prospects for major private investment programmes. The private developers came along just as the federal-provincial work was being wound up. Even so, Alberta budgeted over $87,000 for 1930-31 and $72,000 for 1931-32 for the Research Council. There was haste in the province's bowing out. Having just gained jurisdiction over natural resources, Alberta was being offered a fine chance by private enterprise to bring to completion its development strategy for the oil sands.[45]

* * *

The dehydrator at the Clearwater River Plant was considered to be unreliable. Nonetheless, bitumen produced there had only 1 percent water content. Provincial Archives of Alberta

The University of Alberta and Research Council research was significant because it developed the most promising method of obtaining a useful raw material from the tar sands, which pointed the way to their commercial development. The Alberta researchers, above all Karl Clark, had developed one method of extracting bitumen into an effective but not necessarily an efficient technique. The technique was not novel, but the researchers' achievement is worth noting both for its distinction and its limitations.

The work of Clark and his associates was remarkable for its goal of sponsoring commercial development. In this regard, it would appear that he was motivated less by disinterested and abstract goals of solving an engineering problem than by the promotional and material goals of solving a commercial problem through engineering expertise. Not long-term but short-term goals stimulated his approach and work. Expectation of commercial success led Clark and Tory into often hasty judgements about the success of Clark's work. This haste can be seen as much in 1930 as in 1921 or 1922. Some might enquire if the focus on commercial development slightly warped research, with the result that some preliminary questions about the properties of the bitumen remained unsolved too long, although no promising leads were actually ignored. The possible uses of the bitumen should surely have been examined more seriously much earlier than 1926-27, however, if only to remove misconceptions. For instance, only in 1926 was a serious but even then preliminary industrial analysis made of raw bitumen. The fact that there may be reservations about Clark's work also raises some doubt about Tory's argument concerning the greater benefits of publicly-funded basic research, compared to the more tempestuous approaches of private developers.

These questions aside, the impressive achievement of Clark as well as its general utility must be noted. His achievement is particularly striking when the extent of his funding is taken into consideration. While his own salary of $3,500, rising to $4,000, was more than comfortable, he did not have large capital to spend on research. A rough total of spending on the oil sands research, including all facets of Clark's work, excluding salaries, would be about $88,000. The Mines Branch estimated that the Research Council spent only $75,000, again excepting salaries. Since three separation plants were built and many costs were incurred over the decade, this expenditure was certainly modest. In comparison, Ells's various works, from 1913 to 1930, cost the Mines Branch much more. His cumulative salary was $40,000 and he spent some $85,000 for field research and other experimental work. But Ells was noted as a profligate spender on his junkets.[46]

To summarize: For between $160,000 and $170,000 (plus salaries for Clark and Ells and assorted associates), the two levels of government provided the crucial studies of mining, separation, and paving which would encourage a decade of private development and, later, public and private interests. All the work of these researchers in the twenties was made available freely to commercial interests. While such assistance did not guarantee any commercial success (state regulation and business-government co-operation were not highly developed in the resource field in the twenties), state-sponsored research pointed private developers to the correct methods, as well as warning of the many problems of developing the resource.

One final point might be noted about the research up to 1930. A number of individuals, including Clark's assistants at the Research Council—W.G.Jewett in 1922 and 1923, Sidney Blair from 1924 to 1925, and J.G. Knighton in 1924 and 1925—were young engineering graduates who would become involved again in the oil sands years later. Blair, as a respected international oil consultant, would interpret the commercial feasibility of the oil sands as of 1951; Jewett and Knighton would sponsor research in the thirties and more work in the forties as officials of the Consolidated Mining and Smelting Co. The tar sands clearly held out attractive prospects which captured the interest of young scientists. The "breakthroughs" of the twenties and the reserves locked into the oil sands lured sensible, capable engineers as well as fortune-seekers to work in tar sands development.

The experience of the researchers in the twenties showed the legitimate possibilities of the bituminous sands and the importance of state-assisted research. It also uncovered the

There was some time for relaxation at the Clearwater River Plant as this photograph taken outside the cookhouse shows.

Provincial Archives of Alberta

| Table 2.3: | Research Council of Alberta Spending on "Road Materials Division" 1921-1930 |

year:	amount: ($)
1921	300
1922	3,400
1923	6,000
1924	20,200
1925	15,300
1926	3,450
1927	2,599
1928	2,250
1929	15,000
1930	20,310

N.B. This excludes salaries for Karl Clark, which moved from $3,500 to over $4,000 per annum, and those of various seasonal and short-term assistants.

many difficulties in realizing these possibilities. The years of provincially-sponsored research were—and remain—highly instructive. Funding industrial research did not directly "pay off" in the development of a new industry in the case of the oil sands. But it must be noted that similar provincially-assisted industrial research on coal and natural gas was also conducted in the twenties, with what results our historians have yet to inform us.

The oil sands were not to be a resource developed for purely local needs: thus, even the road-building value of the oil sands was discounted. The oil sands were going to be developed only if the market for petroleum products, and the world conditions which determined that market, were favourable. Yet in making that point clear, the Research Council was teaching a useful lesson. Like the wheat economy, which was then the basic industry of the prairies, so too the new resource economy was prey to international economic factors, no matter what the provincial government tried to do.

III: Promotions and Experiments on the Athabasca: Private Enterprises, 1906-1941

Private businessmen were attempting to exploit the Athabasca oil sands at the same time as government research scientists, but their activities are difficult to trace. Few of them left evidence of their work and even those whose records are available have seldom provided satisfactory documentation of their activities. Most of the private enterprisers' operations were insignificant to the commercial or technical development of the tar sands, however noteworthy they were as examples of frontier developers and the frontier pattern of development. The surviving vestiges stand as a salutary reminder of the intractable engineering and economic problems posed by the oil sands, although they do point to the entrepreneurs' hopefulness and their frustration.

International Bitumen and Abasand Oils were the only operations whose activities are not totally obscured, but they were remarkably different organizations. International Bitumen was the creation of an independent developer, Robert Fitzsimmons, a tenacious businessman who failed only after a long struggle. Abasand Oils was the creation of Max Ball, the epitome of the professional oilman-businessman, but Abasand did not succeed either.

* * *

Three types of enterprises were created to exploit the oil sands. The first, drilling, was conducted by oilmen seeking oil pools beneath the tar sands. The second, paving operations and experiments, was carried out by businessmen seeking to take advantage of the direct road-surfacing use of tar sands on the dirt roads of the prairies. The third, extraction experiments, was undertaken by those who accepted the premise that the bitumen had to be extracted from the oil sands and that it provided a source of crude oil. Drilling, paving and extraction operations were developed in chronological sequence between 1906 and 1941. But only in the 1930s did the two major separation and refining operations of Robert Fitzsimmons and Max Ball begin to take up where Clark's Clearwater plant left off.

In the first phase of activity by private enterprise, a number of speculators drilled for an elusive and illusory "pool of bitumen" below the oil sands. The earliest and most famous of these drillers was the eccentric German speculator, Alfred von Hammerstein, who was active in the Athabasca area beginning in 1906. He eventually gained control of about twelve thousand acres of freehold lands on which he drilled for oil. Hammerstein and his partners—his company, Athabaska Oil and Asphalt Company, was organized only in 1910—conducted four seasons of drilling between 1906 and 1909, which probably resulted in six wells. Much later, in 1924, a forty-seven-foot well was drilled.[1] Hammerstein's early efforts were unsuccessful although they were significant, for he pushed his cable-tool rig to drill 500, 600 and even 1,000 feet into the ground, or so he claimed. He was most notable as a promoter, building speculative hopes for the region, and he went to some lengths to promote his holdings, even attempting to spread false information about drilling results. One informant told the Royal Canadian Mounted Police that Hammerstein went so far as to pour tar into one of his wells prior to a visitor's investigation of it. His work, however, neither benefitted his personal holdings nor the reputation of the oil sands. Hammerstein ended up obsessed by the value of his freeholdings but quite poor.[2]

Hammerstein's efforts were followed by those of other ventures such as the Northern Alberta Exploration Company. This firm drilled six shallow wells near the confluence of the Horse and Athabasca Rivers in 1913 and 1914. After they drilled through rich tar sands layers salt was found to be the likeliest commercial product.[3]

Hammerstein's promotions and other activities before World

War I resulted in a minor flurry of interest in the Athabasca region and the village of Fort McMurray, one of the many speculative fevers which plagued the development of the prairie provinces in that pre-war period. Hammerstein and other land speculators stimulated a provincial government agreement to guarantee loans for an American promoter who proposed to build a railroad from Edmonton to Fort McMurray. The infamous Alberta and Great Waterways Railway was designed to tap both the resource wealth of the Athabasca area and the water transportation links to the entire Mackenzie River basin, a region regarded by even respectable opinion in the late nineteenth and early twentieth centuries as offering fabulous commercial possibilities. The Alberta and Great Waterways line, however, was noted chiefly for precipitating one of the major public finance scandals in Alberta's history. Highly extravagant government loan guarantees—$20,000 per mile for 350 miles—enabled promoters to pocket tidy sums of money. The provincial government of A.C. Rutherford was probably more gullible than culpable, but Rutherford was forced to resign the premiership in 1910. The Great Waterways railway was eventually built. Although construction stalled at Lac la Biche in 1917, less than halfway to Waterways-McMurray, in 1921 the railway reached Waterways.[4]

The discreditable speculations encouraged the Dominion government in 1913 to close the oil sands to further freehold grants. Not until after World War I did the federal government again allow exploration in the area—however, from then on, exploration was on a lease-hold basis.

Of other wells drilled in the area before the twenties, some, such as one drilled in 1915 by J.D. Tait of Vancouver, reached 1,000 feet, while others, like the single well of the Athabasca-Spokane Company, drilled in 1917, went down only 100 feet.[5]

Such endeavours generally had ended by about 1918, perhaps because one recent operation had proved fruitless. In 1914, Imperial Oil, a company knowledgeable and innovative in northern oil exploration—it was to strike oil at Norman Wells on the Mackenzie River in 1920—sent T.O. Bosworth, a renowned geologist, to report on the oil sands. Bosworth estimated that there were 30 billion barrels of oil in place, an assessment which led the North West Company, an Imperial Oil subsidiary, to drill two wells in 1917 and 1918. Nothing of substance was found.

Some well-drilling continued in the 1920s, the most notable being that of the Alcan Oil Company which became the International Bitumen Company in 1927, under the control of Robert Fitzsimmons. Alcan-International was as unsuccessful as any other firm, but Fitzsimmons's tenacity eventually led it into its persistent extraction efforts.[6]

A second group of oil sands developers attempted to find uses for the oil sands as a road surfacing material. This they tried to do by using raw bituminous sands or by heating and mixing oil sand with gravels. These efforts, like the drillings, mirrored the attempts of government researchers. Ells and Clark had attempted to use the oil sands as a paving surface, but they had found the use of raw or mechanically treated sands to be unimpressive. Private developers faced similar results.[7] (See Table 3.1 on mining results.)

Thomas Draper, an oil equipment manufacturer from Petrolia, Ontario, was one of the most persistent experimenters in the use of oil sands for road surfacing, and his results were among the most significant. Draper's experience in oil sands technology was multi-facetted, however, and he was equally active in mining and

Thomas Draper's camp near Fort McMurray, 1927.
University of Alberta Archives

extraction research in the twenties, experiments which were conducted through his company, McMurray Asphaltum and Oil.[8]

As a result of his mining activities, he was able to act as supplier of oil sands to various investigators, including the Research Council, to which he shipped 22 tons in 1922. He mined and shipped a further 500 tons for the Council's Dunvegan yards operation in 1924 and 1925. In 1926, he mined more than 1,000 tons. Thereafter, Draper's mining activities, which had been valuable to researchers as well as being profitable to him, wound down until they ceased in 1936.[9]

Concurrently with his mining activities, Draper experimented with a separation process. This research had commenced in Petrolia in 1920, when he had used a distillation technique in which the oil sand was heated to recover the bitumen. The next year, Draper continued his experiments, attempting separation by using the water flotation method. He built a small four-ton plant at his operating headquarters near Fort McMurray which cost $35,000 to construct and operate, but the structure was destroyed by fire in 1924, and Draper withdrew from separation research. Indeed, the plant had failed to impress Karl Clark; Draper was deeply injured by reports of Clark's criticisms which he considered to be a premature judgement of his experimental process, but he harboured no grudge against Clark once he had abandoned the venture.[10]

Draper's research with oil sands as a paving material, either untreated or mixed with asphalt, began in 1925, and he spent several years publicizing and exhibiting his paving experiments, notably at the Edmonton Exhibition. The most important of his road paving contracts came from the federal Department of Public Works which hired him to pave a small portion of Wellington Street, Ottawa, in 1925, despite the objections of Mines Department officials. In 1930, he was hired once again, this time to pave part of Parliament Hill. Besides taking a number of road and sidewalk paving contracts in towns like Vegreville and Medicine Hat, Draper's most notable paving assignment also came in 1930, when he surfaced twenty-two blocks of sidewalks in Camrose.[11] This work was evaluated by the Research Council's Karl Clark, who appears to have provided the most thoughtful assessment of Draper's success and problems in the paving business. Clark noted that Draper's chief problem was cost: Camrose was charged $1.00 per square yard, a much greater amount than the 66 cents per square yard which would have been charged by Edmonton City Engineering Department for the same job. Even at the amount Draper charged, Clark thought that the operator could not have met his costs. Clark's evaluation of the job also left the impression that the surfacing job was acceptable, although the paving was not remarkably strong.[12]

After 1930, Draper began to withdraw from significant activity in the oil sands (see Table 3.1). No doubt the combination of world economic depression and his own company's financial difficulties were too much for him. He had attempted to demonstrate the possible commercial value of the oil sands as a paving material, but he had failed to provide any convincing evidence that his work merited further government-associated trials.[13]

The third group of oil sands developers experimented in extracting crude bitumen from the oil sands. Their works were the most extensive and, likely, costly ones concentrated in the twenties and thirties. They ranged from highly speculative, extravagant applications of engineering theory to separation by amateur inventors to professional petroleum engineers' efforts at perfecting processes, in a fashion similar to Karl Clark's. The processes can be distinguished by the three approaches identified by Sidney Ells in his Mellon Institute studies. One tried to distill bitumen by applying heat or steam to the sands, then recovering bitumen from the heat or steam mixture; the second attempted to dissolve bitumen by applying a liquid petroleum solvent, then recovering the bitumen from the solvent mixture; the third aimed to separate the bitumen through mixing oil sands with water and then reprocessing the oil-and-water mixture. Only the third method led to extensive commercial experiments. About fifteen or sixteen oil sands separation ventures can be positively identified as having resulted at least in experimental work aimed at using the Athabasca raw material.[14] Unfortunately, most of these firms have not left documentary evidence of their activities; indeed, several were probably mere drawing-board experiments.

Perhaps the earliest work was suggested by D. Diver, whose Diver Extraction Process was attempted in 1920. Diver attempted to ignite oil sands from wells dug into the deposits. His process

Table 3.1: Bituminous Sands Mining, 1924-37

Year	Tonnage	Estimated Value	Price per Ton
up to 1924	531	$ 2127	$4.01
1925	1148	4594	4.00
1926	528	2112	4.00
1927	2706	10824	4.00
1928	94	374	3.98
1929	989	3956	4.00
1930	2067	8268	4.00
1931	1015	4060	4.00
1932	343	1372	4.00
1933	466	1662	3.57
1934	862	3449	4.00
1935	40	160	4.00
1936	—	—	—
1937*	35	142	4.06

*Last year reports published

Source: Calculated from Canada, Department of Mines, *Mineral Statistics of Canada*, 1924 to 1937.

involved injecting a gas-powered flame into the well, which was intended to melt pools of bitumen large enough to be pumped to the surface. Although a notable effort, the Diver Process did not work.[15]

In addition to the Draper experiment already described, at least two other distillation processes were devised. One Calgary venture, the Georgeson Extraction Process, was quite extensive, according to photographic evidence. The process was an *in situ* application of steam into oil sands shafts. The Georgeson workers dug three shafts of 65, 55 and 50 feet in 1924. Then steam was applied at a pressure of 90-lbs. per inch for some 12 hours. When these wells were uncapped, a mixture of water and oil was usually expelled, but no uniformity or continuity of separation was achieved apparently.[16]

The most promoted of the distillation schemes was that of Major-General Bertram Lindsay. Lindsay's apparently solid

Absher attempted to extract bitumen from the oil sands by injecting steam into the wells.   Provincial Archives of Alberta

Jacob Absher's camp at Saline Creek in 1930.   Provincial Archives of Alberta

Absher's experiments with steam injection were unsuccessful.

Provincial Archives of Alberta

67

finances—his organization had a shale-oil extraction operation in Wales which, Ells thought, indicated a $400,000 investment—and his eagerness were most inviting to Federal officials in 1920. Lindsay proposed to extract crude by heating the sands and condensing crude bitumen. Henry M. Tory's skepticism in the face of the Lindsay proposal was sensible, for the Lindsay group did not get beyond initial study of the oil sands, and there is no evidence that they spent any money there.[17]

By the mid-twenties, a number of privately-funded experimenters were examining variations of the water mixing and separating technique. These researchers included the former oil sands driller operator, James Tait. In the early twenties, having failed to extract bitumen by drilling, Tait proposed to develop a hot-water wash and flotation-cell separation system in order to extract bitumen. This proposal did not get beyond the blueprints,[18] but this was not the case with the work of English industrial chemist Ernest Fyleman, which resulted in both patents and early publication in 1922, to Karl Clark's chagrin. Fyleman, however, does not appear to have been interested in actual development of his process.[19]

A remarkable experimenter was Jacob Absher, a Montana oil-man, who tried two innovative separation processes in the late twenties. Backed by the Bituminous Sands Extraction Co., whose principals were William Fisher of Calgary and Curt Smith of Wetaskiwin, the indefatigable Absher worked for four years with the conviction there were six billion barrels of oil in the McMurray region. Absher first attempted to separate bitumen by injecting steam into shallow twenty or thirty foot wells. He continually had problems selecting well-sites and often missed the underlying tar sands deposits. The steam-injection pipes also clogged up so that he could not drain bitumen separated by the steam. Later Absher tried to inject burning kerosene into the shallow wells but this effort also failed because the pipes melted.[20] A bemused Karl Clark observed Absher's efforts each summer, admiring the innovative approach but regretting the amateur methods. Like many other pioneers, Absher did not work in a systematic fashion, although Clark had tried to convince him otherwise. Despite his ingenuity, Absher's results remain obscure. He eventually concluded that his processes were most amenable to the heavy oil of the Lloydminster-

Wainwright area and he abandoned the Athabasca country.[21]

One aggressive promoter of the water extraction process was the Toronto mining engineer, William P. Hinton. In 1929, Hinton had impressed Alberta government officials like Trade Commissioner Howard Stutchberry, who was an inveterate booster of the oil sands to business interests. Hinton claimed to have a ready solution to the extraction problem, no doubt relying on his experience with mining-type ore dressing technology. Hinton received only guarded co-operation from Karl Clark, who helped the Toronto engineer to procure oil sands samples. Clark warned Hinton that his ideas were far removed from the experience of the Research Council, which Clark explained in detail.[22]

Hinton argued that the "Laughlin centrifuge," an ore-dressing machine, would work on the oil sands. His group operated an experimental unit at the Dunvegan Yards in 1929. The process involved mixing quantities of kerosene to the sands, then adding hot water. The resulting liquid was injected into a centrifuge where the sand was separated from the oil-water mixture mechanically. Karl Clark reported that the so-called crude oil resulting from this operation contained up to 25 percent kerosene. He wondered whether much or all of the kerosene was separated from the liquid, let alone the bitumen. Finally, he noted that the centrifuge was highly wasteful of steam.[23]

Hinton was not encouraged sufficiently by results to continue the experiments. His own engineer felt that the "extreme difficulty" in separating bitumen was caused not by any problem with the centrifuge, but only with "conditions." He used only 5 of the 150 tons of oil sands purchased from Tom Draper before abandoning work. Although Hinton secured one of the first two oil sands leases issued in 1930, he later gave up his interests. In the end, as Henry Marshall Tory observed to R.C. Wallace, Hinton's machinations and promotions succeeded only in getting him an exploration lease and in giving the tar sands another "black eye."[24]

* * *

All other private ventures pale in comparison with the two major undertakings begun in the late twenties. Robert Fitzsimmons and his International Bitumen Company present a study in an obsessive, amateur, and woefully undercapitalized commercial

Robert C. Fitzsimmons was the last and most tenacious driller in the Athabasca area. Provincial Archives of Alberta

experiment. In contrast, Max Ball's and Abasand Oils Limited was systematic and professional. Yet each operation failed to gain commercial success, although for different reasons. The stories of these operations reveal evidence of the problems of developing commercial oil production—as was their ultimate fusion into one company in the fifties.

Robert Fitzsimmons, a Prince Edward Island-born promoter, was the last and most tenacious driller in the Athabasca area. After several years in the United States, where he acquired real estate holdings in the Spokane, Washington area, Fitzsimmons moved to Edmonton to take over the Alcan Oil Company in 1925. In 1927 he reconstituted the firm as International Bitumen Co.[25]

As late as 1929 and 1930, he continued to drill in the vicinity of Bitumount, Alberta (a postal address he gained approval for in 1937) on leases acquired in the early twenties. His persistence was Fitzsimmons's characteristic trait. He continued in his drilling operations to "find" bitumen pools, the discovery of which he announced at regular intervals in the late twenties. Skeptical Mines Branch officials like H.H. Rowatt accepted that Fitzsimmons was serious in his work but rejected the veracity of his interpretation of drilling results. In fact, there is a simple technical explanation to Fitzsimmons's periodic "gushers" of crude oil. The heat of the drilling-bit on occasion would act like any other heat treatment and would lead to a limited separation of bitumen from the sands, although there were no more than a few coagulated lumps.[26]

In February 1930, however, Fitzsimmons abandoned a well at 260 feet. After some weeks of rather uncharacteristic inactivity, he began to construct the first of several small hot-water separation plants on his leases. By June 1930, Fitzsimmons was running a tiny hot water separation plant. This plant, a curious Karl Clark reported, involved shovelling oil sand into a tank and mixing it with hot water. The mixture was then transferred to a second tank, where the froth of the water and bitumen was skimmed off by hand. Sand tailings were shovelled out when the plant was shut down. Clark remarked that this primitive operation resulted from Fitzsimmons's previous observations of his own plant on the Clearwater River. Primitive the small plant was, for Sidney Ells reported that the seven-man crew produced about 300 barrels of crude bitumen between 12 June and 25 August of 1930.[27]

The origin of Fitzsimmons's first plant is worth pursuing briefly. In his various *apologia* Fitzsimmons claimed that he had devised his separation operation himself, although it is clear that the plant was built only after Fitzsimmons observed the Research Council operation. It even added new features, like a pug-mill in 1931, which followed the Research Council's modifications done in 1930. Max Ball later remarked that Fitzsimmons had done a very creditable job of copying Clark's operation. Later still, Fitzsimmons referred one of his plant superintendents to Clark's former millwright, John Sutherland, when operating problems occurred in 1937. That this advice was available indicates a longstanding co-operation. Little is remarkable in this except that it indicates that Fitzsimmons was not original in devising his plant operation, despite his claims to the contrary (sometimes taken at face value by recent writers). It can be concluded that Robert Fitzsimmons was a clever adapter of the Research Council work to his credit as well as Clark's.[28]

The original International Bitumen operation was rebuilt in 1931 with a salvaged boiler to provide heat for the plant. No reports on production for the year have survived (if they were ever made), but the plant ran well enough to impress Karl Clark. Clark described the plant operation in detail:

> It is the same type of plant we had on the Clearwater River. The sand is heated and mixed with water in a pug mill and discharged into a second pug mill filled with hot water. The bitumen is skimmed from the surface of the water into a reservoir. This bitumen is next put into a third pug mill and stirred around in cold water. This operation removes clay and some water from the crude bitumen. The bitumen is then passed on to a dehydrator where the water is boiled out of it by heating with steam coils. It is then run to a final storage and settling tank.

The peculiar qualities of Fitzsimmons's supply of non-acidic tar sand, which was also extremely rich in bitumen (about 15 percent by volume), meant that it was easy to process. There was no need to use the silicate of soda emulsifying agent that Clark was forced to use.[29] There were no problems with erratic rocks or solidifying sands.

If Clark was impressed, Fitzsimmons was exultant. In late 1931, as economic times worsened everywhere, he tried to promote his International Bitumen to his shareholders and his products to Albertans. He pointed to the presence of oil sands with 30 percent bitumen only two feet under the ground of his leaseholds, a slight exaggeration. He also hinted at a discovery of a rich vein of pure bitumen—he was an incurable promoter! Finally, he referred to his "invention" of a remarkable process to refine bitumen from the rich sands. This separated bitumen Fitzsimmons hoped to sell to users of asphalts for roofing and road surfacing.[30] In the worsening depression, International Bitumen was unable to create any markets for its potential product, let alone to find a large pool of capital to hasten

Fitzsimmons attempted to locate bitumen "pools" by drilling with equipment like this.
Provincial Archives of Alberta

development. Edmonton's City Engineer A.W. Haddow, a possible buyer, observed wistfully that Fitzsimmons had only to prove the technical merits of his product and Edmonton would buy large quantities of bitumen. Fitzsimmons never met this test and Haddow had too much experience in the previous decade's paving experiments to accept on faith any claims for oil sands products.[31] By 1932, Fitzsimmons claimed to have spent about $200,000 on development ($100,000 in the 1930 and 1931 seasons, for he had previously claimed to have spent $100,000 up to 1930). But he had no more capital reserves to call upon to continue production or to expand his little plant. Consequently, his plant did not operate from 1932 to 1937.[32]

The heat of the drilling bit would occasionally separate bitumen from the oil sands, as this photograph taken at the International Bitumen camp shows.
Provincial Archives of Alberta

International Bitumen had a stock of 144 barrels of bitumen in March 1933. But Fitzsimmons noted that he had no buyers for even this small inventory. He stated that he had tried to sell his bitumen for roofing tar but had not succeeded in getting any large orders. He had shipped Max Ball a supply of the rich Bitumount oil sands (Ball was conducting laboratory tests of his own process in Denver and Toronto at that time). But the shoe-string International Bitumen, chronically short of funds and built on a tiny scale (even his later improvements left the plant with but 100 tons per day capacity, only double that of Clark's Clearwater prototype) meant that Fitzsimmons had few prospects. Ball's Abasand Oils project was the respectably engineered and capitalized operation, the one garnering the publicity which salesman Fitzsimmons craved.[33]

Through his herculean promotions (Fitzsimmons drummed up some stock sales in tours of the mid-western United States and even Great Britain in the mid-thirties) and, it seems, his own other endeavours in real estate, Fitzsimmons managed to pull together enough capital for International Bitumen to plan for expansion and operation in 1937. Throughout this period and later, Fitzsimmons looked at his problems as being the result of sabotage by perfidious employees, provincial government officials, or the large oil companies or a combination of the three. Factors such as capital sources, proper markets, or verifiable technical evidence for his claims never disturbed Fitzsimmons's assessment of his firm's value.[34]

In 1936, however, International Bitumen hired a petroleum engineer who had been overcome by Fitzsimmons's contagious fever of belief in the value of the oil sands. Harry Everard was an engineer with Richfield Oils in Los Angeles. He had previously consulted with Fitzsimmons on the possible refined uses of the Athabasca bitumen. Everard began his work just as committed to the great promise of the oil sands' potentials as was Fitz-simmons.[35] This promise was not without truth. Growing activity in the mining industry of the Canadian Shield meant that, from the Athabasca-Peace River system to the Mackenzie River, a considerable market for petroleum products was emerging. Both Abasand Oils and International Bitumen were well placed to take advantage of this market, which offered them the chance to supply the mines of the north. Their locations gave them an

Fitzsimmons's primitive separation plant at Bitumount in 1930.

advantage over southern sources and suppliers, even Imperial Oil's tiny operations at Norman Wells, Northwest Territories. Moreover, International Bitumen had been able to produce a good quality crude bitumen, much to the surprise of both Clark and Ball.

Harry Everard was hired at a salary of $350 monthly in order to build an oil refinery and refabricate the separation plant at Bitumount. International Bitumen could then begin to market a refined petroleum product to the northern mines. Everard began optimistically enough. He sought to build a 700 tons per day separation plant and a refinery with similar capacity that could produce light and heavy diesel oil as well as asphalt. This he thought he could do for 71.8 cents per barrel. which would yield International Bitumen the return of $5.42 a ton (or $3,793.91 per day of operation). Everard and Fitzsimmons relied on the captive market in the north, particularly the mines owned by Consolidated Mining and Smelting (CM&S), which had extensive operations on Lake Athabasca and Great Bear Lake. The two million gallon market held a very appealing sales potential to the hopeful Fitzsimmons.[36]

The reality of rebuilding and running a plant, using second-hand equipment, that improvisation being demanded by International

Bitumen's continuing shortage of funds, proved more difficult than Everard had thought. In October 1936, after weeks of effort in getting operations started, Everard and the plant engineer, Victor Engman, reported on the disastrous results—up to 50 percent sand and water content marred the separation process. This extremely raw crude was actually run through the refinery, with the result that the refinery could not "cut" or refine the crude. Engman asked if the separation plant could be redesigned, and Everard requested Fitzsimmons for further funding in order to do this.[37]

After a winter lay-off, a forty-six-man crew began rebuilding the separation plant in the spring of 1937. On 25 June, Everard wrote to Fitzsimmons that the separating plant was producing a good quality bitumen. This crude tested to 1.2 percent sand and 15 to 20 percent water content. Between 15 and 34 barrels of bitumen were produced during each hour of operation. Whether the plant could run for more than a few consecutive hours, Everard didn't say. The refinery could handle up to 450 barrels of crude per day, at least as long as the water content of the crude was below 20 percent. Everard was pleased. Louis Romanet, Max Ball's sales representative in the Athabasca region, visited the plant site and confirmed the story of International Bitumen's successful start-up. At last, Fitzsimmons had managed to interest northern mining firms with his potential fuel oils; while he had no agreement as yet, International Bitumen's fortunes appeared to be very much improved.[38]

From this point of success, however, conditions at Bitumount began to deteriorate. Breakdown after breakdown in the separation plant interrupted the summer's work. Necessary equipment either arrived late or failed to appear. Nonetheless, from time to time the separation plant worked steadily. On two eight-hour shifts per day, it produced 200 barrels of crude at the very best. Because of the limited separation result, the refinery worked at about one-third capacity. On 8 September 1937, Everard wired collect to Fitzsimmons that he was suspending plant operations and returning to Edmonton. Impatient with the engineering difficulties and probably annoyed at friction with and among the plant workers, Everard admitted defeat.[39]

In his later explanations, Everard cited exasperation with Fitzsimmons's failure to pay either his salary or those of the rest of the crew. Everard spent several months, until April 1938, negotiating with Fitzsimmons over his own claims on the firm. At one point he agreed to try to sell company stock in order to raise funds that he might place against his wage claims, but nothing came of this. After long delays, Everard continued negotiations with Fitzsimmons through a lawyer. It was at this point that Fitzsimmons hired his own lawyer and began to denounce Everard as a saboteur. More than a year and a half after Everard left his employ, Fitzsimmons charged that his chief engineer had induced the men to walk off the job in early September of 1937 and that he had sabotaged the design of the separation plant and refinery.[40]

Such charges and counter-charges are worth examining, for they were serious in themselves and they provide an explanation for the plant's failure in 1937. The greatest significance of the Fitzsimmons-Everard feud lies in the form it took, which closely resembled that of others in which Fitzsimmons engaged. For example, Fitzsimmons's one-time head driller and later plant mechanic, Frank Badura, faced similar recrimination when he claimed wages dating back to the early thirties. Fitzsimmons charged that Badura had virtually wrecked International Bitumen's prospects from 1935 to 1937. The persistent Badura kept asking for his wages and his wife's and in the end won at least partial payment. But it was not until June 1943 that Frank and Marie Badura received $1,668 in back wages. As with Badura, so with Everard, although Everard received only the tiny sum of $300 in 1943 in return for an agreement to drop further claims against Fitzsimmons. Both claims were settled only after Fitzsimmons had sold International Bitumen to Montreal businessman, Lloyd Champion.[41] Whether or not Everard was incompetent cannot be settled here, although it might be noted that he had supervised some production, despite Fitzsimmons's charges that he had not done even that. In sum, the unhappy 1937 season indicated that International Bitumen was neither financed nor designed on solid enough foundations. In his disappointment with these weaknesses, Fitzsimmons reacted with understandable disappointment as well as considerable vindictiveness.

In October 1937, Elmer Adkins, a Medicine Hat-born recent engineering graduate from the University of Alberta, was

The Separation Plant of International Bitumen at Bitumount, 1935 or 1936.

Provincial Archives of Alberta

appointed International Bitumen's superintendent, replacing Everard. The young Adkins had spent the previous season working as the plant's chemist, at $150 per month, a salary which was not raised despite his exalted new title. Before that, Adkins had worked at Max Ball's Abasands plant, experience which undoubtedly helped him at Bitumount. Adkins accepted the challenge of trying to get the plant in operation. By this time, Fitzsimmons had signed a contract with CM&S to supply seventy thousand gallons of fuel oil for its northern Saskatchewan mines in 1938. With such a boost, International Bitumen was able to move once again towards separating bitumen and refining fuel oils.[42]

To his great credit, Adkins worked with his crew through 1938 and, by that summer, International Bitumen actually was producing and refining oil. Despite late or missing pay cheques, despite late or missing supplies, the plant had been extensively rebuilt between January and June 1938. The inexperienced crew was trained to run the separation plant and the refinery; a steam engineer was hired to operate the uncertified steam boiler (this was inspected and repaired only after Adkins begged Fitzsimmons to take action). Records are sketchy, but it appears that

The Refinery of International Bitumen at Bitumount, 1936 or 1937.
Provincial Archives of Alberta

Adkins managed to grind out sufficient production from the small plant on the Athabasca River that International Bitumen met some although not all of its contract with CM&S. International Bitumen earned over $3,000 for supplies sold to CM&S in 1938.[43]

But this success did not mean that International Bitumen was at last a viable company. Fitzsimmons had had to scramble to obtain money for 1938. In January, he had written his treasurer, Robert MacMicking, from Chicago that he had turned up $3,000, with further monies assured. Indeed, quite the opposite was true. Elmer Adkins resigned his position on 6 September 1938, partly because he had received another job offer but also because he had seldom been paid for his work at International Bitumen. "Please send me some money," he plaintively requested Robert MacMicking. He needed to clear up some debts in Edmonton and then to move to Brandon, where he was taking up a post with the Anglo-Canadian Oils refinery. Adkins continued to request back pay for several years; he appears never to have received one cent but, then, he did not revert to the courts for redress.[44]

The most remarkable aspect of his continued dealings is that Adkins remained so sympathetic with Fitzsimmons. This indicated his abiding commitment to the possibilities of the oil sands. Even as Adkins became a highly successful petroleum engineer and businessman (he became a senior executive with Domtar in later years), he kept his contacts with oil sands developers. In 1941, while he was working for the Born Engineering Company, a large American petroleum engineering contractor, Adkins discussed with Fitzsimmons the possibility of building a large-scale oil sands plant. Adkins returned temporarily to the oil sands in 1948 and 1949 as Superintendent of the Alberta government's splendid experimental operation at Bitumount which was built by Born Engineering Company. Adkins's relations with Fitzsimmons and his interest in the oil sands, like that of Fitzsimmons, Karl Clark, Sidney Ells, Sidney Blair and others, showed how intriguing if not hypnotic the oil sands were.[45]

International Bitumen was the first commercial separation and refining operation in the oil sands and it was established despite many problems. Therefore, the company's 1938 operations deserve closer examination both as financial and technical undertakings.

In the 1938 season, Adkins employed a crew which varied in number from 14 to 40, although when production was undertaken it employed about 30 men. The monthly payroll varied from about $250 in February to more than $1,800 in August. The men were paid monthly, with the senior tradesman, the stillman, and mechanic receiving $75, while labourers were paid from $35 to $50. The costs of the men's lodgings and meals, to a maximum of $19.50 monthly per person, as well as "medical fees" (probably workmen's compensation) of 78 cents monthly, were deducted from their earnings. The plant kept men on payroll from January until early December 1938, when the last sixteen walked off the site.[46]

This season of production actually allowed International Bitumen to report revenues from sales. Robert MacMicking advised Fitzsimmons of the following income from sales[47]:

CM&S (fuel oil sales)	$ 3,265.50
Edmonton (city)	724.03
Wirtz (unidentified)	2,237.37
Dominion Government (Misc.)	28.00
(450 tons asphalt)	14,040.00
Total	$20,294.90

Against these revenues, International Bitumen had to face wages of as much as $14,000 (data are available for only seven months, which prorates that amount to about $1,160 per month). Salaries for Fitzsimmons and MacMicking are not included, although MacMicking took $1,430.86 in salary payments in 1938. He was owed $6,569.14 when he died in 1941. Unfortunately, no solid information about Fitzsimmons's salary exists. MacMicking hoped that the firm's solicitor, Charles Becker of Edmonton, would not pay the men any wages in December 1938, for such payments would totally bankrupt the business, he warned Fitzsimmons. There was also the matter of employees' back wages. Thus, despite the remarkable achievement of International Bitumen in production and income for 1938, the company was unable to pay its costs beyond basic operating expenses, indeed most of the men do not seem to have been paid.

Bitumen produced at the International Bitumen plant was barged away to markets. Provincial Archives of Alberta

Fitzsimmons left for Chicago late in 1938 and remained there for the next two years, partly to avoid creditors, but also to drum up financial backing. Only in mid 1940 did MacMicking advise Fitzsimmons that his creditors were sufficiently exhausted that he could safely return to Edmonton, which he did.

The major result of 1938, then, was that International Bitumen was insolvent and would remain so until it was sold to Montreal financier Lloyd Champion in 1942. The attempt to press into production had effectively killed Bob Fitzsimmons's International Bitumen.[48]

Two engineering assessments of the plant were written in 1938 and these help to explain why there were no further attempts to resurrect International Bitumen. Each makes clear that International Bitumen was an unreliable operation because, as one engineer reported, its design was "inefficient and haywire." The plant had been built in the cheapest possible manner, according to G.J. Knighton of CM&S. He had toured Bitumount before Adkins's resignation, commenting on the plant's ability to process even 100 tons of oil sand per day and to produce between 250 to 700 barrels of bitumen. Knighton found the production of "so-called fuel" oil from the separated bitumen to be amazing, given the shortcomings of the plant's equipment. He

While the employees at International Bitumen were not well paid, Robert Fitzsimmons certainly did not live luxuriously himself. This photograph of his cabin at Bitumount was taken in the 1930s.

Provincial Archives of Alberta

noted that the separation plant's claimed recovery rate of 75 percent bitumen was probably generous but reflected that a lack of data inhibited any analysis.[49] Similarly, Vancouver engineer John Fairlie reported the problems of assessing the operation, although he stated that the "obsolete and inefficient" machinery and its unsophisticated design made clear that production was limited. Like Knighton, Fairlie noted with amazement that the plant actually yielded a respectable product from its Heath Robinson machinery. Even Sidney Ells, long sympathetic to International Bitumen, concluded by 1939 that the firm had failed due to gross mismanagement and worthless equipment.[50]

Later, in 1943, Lloyd Champion, the new owner of International Bitumen, now renamed Oil Sands Limited, hired an engineer to estimate the cost of producing oil using the company's equipment. L.C. Stevens calculated that crude bitumen could be produced for 76.2 cents a barrel if production achieved 320 barrels daily. He remarked on the general effectiveness of the plant process but commented that improvements were necessary. He estimated transportation costs to Edmonton and concluded that crude bitumen might be sold there for $2.66 a barrel. This was $1.20 below the Calgary market price, he noted. Any bitumen produced from the Bitumount properties would have a good chance of being marketed successfully on the southern Alberta markets, he argued.[51]

Champion was not an obsessive experimenter, but rather a businessman who sought to turn his holding into a profitable property. In any event, Champion was not about to compete with the federal government's experimental work then underway at the former Abasand Oils plant. Thus, under Champion's lead, Oil Sands Limited undertook only limited maintenance and rebuilding work during 1943 and 1944 in order to attract possible partners or buyers. The company never moved toward production again. The properties eventually were merged with the Abasands Oils holdings in the 1950s. Much to Bob Fitzsimmons's disappointment, International Bitumen itself did not re-emerge as the great industrial enterprise he had envisioned.[52] But its successor did in a way do so.

* * *

In 1921, James McClave, a Denver, Colorado, petroleum engineer, examined the properties of Athabasca tar sands. Later, McClave worked in Denver on a prototype hot water separation plant. McClave's early process was developed in 1923 at a cost of $12,500. It involved application to tar sands of water heated to 120 degrees Farenheit mixed with 1 to 1.5 percent bentonite. The resulting wash was sent through a pug mill, where it was ground further and heated. The mixture was then fed into a flotation cell, where the froth of bitumen and water thus generated was skimmed off.[53] This early plant did not appear to have been operated for consecutive years, but its successes encouraged a group of experienced oil men to maintain an interest in the oil sands. The American-trained oil engineers Basil O. Jones and Max W. Ball, prosperous, knowledgeable and venturesome, focussed their attention on the Athabasca oil sands. By 1930, the group, with Ball as its head, was prepared to seek leases and to attempt to construct a commercial-scale oil sands separation plant for the production of petroleum and petroleum products.

The pivotal figure in this operation, Max Ball, was a study in contrast to Bob Fitzsimmons. Born in 1885, Ball was a trained

Max W. Ball, the American-trained oil engineer.

University of Alberta Archives

mining engineer, a graduate of the Colorado School of Mines. He later worked for the United States Geological Survey and then studied law at George Washington University. By the twenties, Ball had obtained varied experience as a public servant and private businessman, as a geologist and as an oil executive. He possessed the type of training, and experience in dealing with both business and government, appropriate for the kind of complex business negotiations that the development of a resource area like the oil sands involved. By all accounts, he was also remarkably phlegmatic and tactful, two qualities which were useful in dealing with the angular personalities of oil sands workers like Fitzsimmons, Ells and, in his way, Clark, who were obsessed by their pioneering work.[54]

Ball and his group immediately impressed Karl Clark with their willingness to observe and to learn from the conclusions of the Research Council's 1930 operations. Ball also impressed federal officials, still administering Alberta's natural resources, with the solidity of his technical knowledge and financial support. Ball applied for oil sands leases, proposing to spend $150,000 developing a commercial separation plant based on McClave's hot water extraction process and the Research Council's findings. The proposal was accepted and Bituminous Sands Permit No. 1 on properties at the Horse River and farther south on the Athabasca were granted to Canadian Northern Oil Company, later called Abasand Oils Ltd.[55]

Regardless of his impressive background and associates, Ball and his firm faced daunting problems. McClave's experimental plant, like Clark's pilot plant, was still only a small-scale operation untested on continuous or extensive operations. Moreover, no obvious market existed for the products of any separation process committed to fuel oils production, rather than road surfacing materials output. Finally, Ball was launching a highly speculative scheme just as the financial crisis which commenced in October 1929 was beginning to have permanent and extremely severe effects, and just as the North American oil industry faced a period of significant decline, with over-capacity and very weak prices marking the conventional industry for most of the next decade.

Despite these imminent problems, Ball proceeded confidently. He worked out an agreement with Canadian and Alberta government officials that his Canadian Northern Oil Company would take up both the work and the equipment of the Research Council of Alberta's Clearwater plant once the Bituminous Sands Advisory Committee's mandate had expired. Late in August 1930, Ball paid the Committee some $9,200 for the Research Council's Clearwater plant.[56]

Ball's engineers began planning to construct a large, 500-ton per day separation plant and to arrange a quarrying system of a similar scale of production. Throughout 1930 and 1931 Ball scrutinized the Research Council plant and records (he had moved to Edmonton), while his associate, James McClave, continued to experiment in the Denver laboratory plant. Even at this time, it appears that Ball was having some problem in arranging financing that would ensure construction on schedule. An initial $22,900 was spent by June 1931, when Ball admitted to Karl Clark that he was engaged in a "desperate struggle" to continue. This struggle was to go on for five years.[57]

The financial setback meant that Canadian Northern Oil Company had to solicit financial support. This in turn meant that the company had to convince potential investors that it had a commercially-efficient method of separation and the markets for its product. Such practical problems, so similar to those of International Bitumen, explain the fact that Ball conducted a lengthy series of experimental operations for the next five years. McClave built a one-ton per day and then a 2.5-ton per day pilot plant at Denver by 1933. Together with McClave, Ball transferred laboratory testing from Denver to Toronto in 1934. There, distilling and refining equipment was added to the tiny separation plant. The operations had been designed through the services of a large American engineering consulting firm, A.J. Smith Company of Kansas City. It was not until 1935 that this work resulted in sufficiently-improved engineering technology that Ball was able to find financial support. At this point, Ball had convinced the brokerage firm of Nesbitt, Thomson to finance construction at the Horse River reserve. Thus, only in late 1935 was Abasand Oils, as the firm was renamed, in a position to build a commercial-scale plant.[58]

Ball realized that supplies of and prices for conventional oil in Alberta precluded direct competition from the oil sands in the established markets at that time. During the period of testing and selling his company's process and potential, he tried to interest the major mining company of western Canada, the CM&S of Trail, in having its northern mining operations supplied with fuel and lubricating oils from Ball's future plant. To the surprise of S.G. Blaylock, the mining company's vice-president in Trail, Karl A. Clark wrote a very positive appraisal of the hot water separation method developed by Ball's firm. Clark also made a favourable estimate of costs of producing crude bitumen from oil sands: Clark thought that a crude oil could be made for under $1.00 a barrel. He emphasized that the separation process was proven and that semi-commercial work would be successful—and dependable. Clark therefore drew CM&S into the orbit of parties interested in the oil sands. Through Clark, in fact, CM&S conducted its own studies of oil sands separation in 1939. Impressed by its findings, the firm supported Ball's work by contracting to purchase fuel oil supplies. For his part, Ball had found the assured local market for his product that he needed to sustain his business.[59] He had done so years before Fitzsimmons nabbed a small share of that northern market for his own tiny concern.

In late summer 1936, the Abasand Oils plant was completed and ready to run at 250 tons per day input capacity. The plant, Mines Branch chemist W.P. Campbell reported, was composed of a pulper, pug mill, quiet-zone flotation cell, diluent mixer, settling

The Abasand Oils Ltd. plant was completed and ready for production in 1936.   University of Alberta Archives

tank (where the bitumen and water mix was skimmed) and an Akins classifier for removal/recirculation of tailings. Campbell thought that operations had been very limited but that a satisfactory beginning had been made. This process contained the one feature, the diluent mixer, which distinguished it from Clark's or Fitzsimmons's work. As Max Ball described the process, the Abasand plant required a two-stage operation:

> Mild abrasion in warm water breaks the films and gives a pulp of water and sand through which are disseminated particles of oil. In a properly designed flotation cell, the oil particles will be picked up by air to form bubbles that float to the surface. The froth thus formed is high in mechanically entrained water and mineral matter, which will not settle out because of the high viscosity and specific gravity of the oil, but which quickly settles out if the oil is diluted with naphtha or kerosene, leaving a clean oil that can be pumped through a pipeline. The diluent can be knocked out in the refinery and returned for re-use.

As Ball made clear, then, the Abasand process combined a solvent extraction with the hot water extraction. Such complex engineering as this double system imposed (combining two of the three possible ways of separating the oil from the sand mixture), was to hinder the Abasand plant as long as it operated. Unfortunately, no production records of the early operations have survived.[60]

On the basis of its initial work, the plant was redesigned to process 400 tons of oil sand per day. Reconstruction was delayed, however, as manpower and materials shortages slowed the renovations. Ball had claimed that operating costs in 1936 had amounted to $65,000.00; by 1937 total construction and operation costs amounted to $371,328.15, and he still faced the costs of building a refinery on the Horse River plant site. To top off matters, the Abasand engineers were finding it almost impossible to deal with the problem of quarrying in the rock-impregnated and particularly hard oil sand of the Horse River. Unlike Fitzsimmons, Ball had not been fortunate in finding an easily-processed raw material. As a result, Abasand was forced to deal with the problem of finding effective strip-mining equipment. It was discovered that the solution was to use a shale-mining planer, a drag-line shovel that was proven in quarry work. Somewhat surprisingly, this equipment was never well-adapted to the hard deposits.[61]

For three years the reconstruction and testing of Abasand was dragged out. Delays were caused by such technical problems as the adaptation of quarrying equipment. In 1939, Ball requested that the federal government, which still controlled the oil sands leases which Ball held under a special dispensation, waive Abasand's royalty bill for three years. In return, Ball undertook to spend $200,000 on upgrading the facilities; this was in addition to the nearly $500,000 which the plant had cost by that date. Even as the plant moved toward production, it was caught in a somewhat difficult enterprise of further improving its plant and bearing the inflating costs. The consolation was that CM&S had contracted to purchase any production of Abasand. As early as 1938, such a contract had been made, although Abasand was not able to begin to deliver on the contract until 1940. Throughout this period of agonizing delay in rebuilding and negotiating for markets, Abasand was concerned lest Imperial Oil suddenly expand production at its Norman Wells oil operations. The erroneous assumption that there was little to produce at the northern wells encouraged Ball to continue during the late thirties.[62]

Finally, in 1941, Abasand began to operate on a regular basis. From 19 May to 21 November, the plant separated bitumen from oil sand and refined a line of fuel oils. Production was quite extensive, with 41,265 gallons of gasoline, 70,700 gallons of diesel oil, 137,550 gallons of fuel oil, and 375,235 gallons of "residuum" produced. The refinery also produced 319 tons of coke. No exact record of the quality of the products exists, although testimony suggested that separation continued to be rather poor at times. Still, the bitumen was made into petroleum products. The process worked well enough that Karl Clark expressed some admiration for it, although he remained skeptical about the need to combine hot water with solvent distillation. He persisted in believing that hot water separation alone was likely to be technically and mechanically superior, and, therefore, economically so too. But all such discussion became academic when the Abasand Oils separation plant burned to the ground on 21 November 1941. The fire reduced to ruins the commercial experiment of Max Ball, James McClave, A.J. Smith and their associates and with that went the approximately $700,000 investment in Abasand.[63]

Abasand had, at the end, operated effectively enough. The faith of its engineers that the process would produce useable petroleum was justified. But the plant had not proven the economic soundness of extraction and refining or the possible scale of oil sands activities. Both questions were becoming increasingly important to the governments of Canada and the United States for, by 1941, Canada was involved in World War II and the diplomatic situation, particularly regarding Japan, was ominous. Thus, the problem of assuring petroleum supplies had become extremely important to Canada and, because of this strategic consideration, the federal mines branch examined the oil sands as a source of supply. However, G.R. Cottrelle, Canada's oil controller, warned Munitions Minister C.D. Howe in February 1942 that Abasand could prove useful commercially or

The Abasand Oils Refinery and Separation Plant began to operate on a regular basis in 1941.   University of Alberta Archives

strategically, only if it drastically expanded its output to 10,000 barrels a day. That would mean a shift of operations to the far more bitumen-laden reserves of the Mildred Lake area, farther north on the Athabasca River.[64] But could Abasand Oils afford a move? And was its means of separation as convincing a technical achievement as Ball and McClave claimed?

Ball himself had recognized in 1941 that a drastic up-scaling of the Abasand operation was necessary. A 10,000 barrel-a-day-plant could run for twenty years at the Mildred Lake lease-site. But that plant would cost $4.5 million to build, to which a $300,000 pipeline to Edmonton and a $3,150,000 refinery there would be added. Ball claimed that this $8 million bill compared well with the $8 million per year that it cost oil companies to squeeze oil from deep Turner Valley deposits, which had come on-stream within the previous three or four years. Turner Valley wells, at this time, required over $200,000 to complete. The experience of Pacific Petroleums indicates that financing well completions was an intermittent affair on any exploratory acreage. Ball suggested that if the economically marginal Turner Valley field was viable, then so was the Athabasca area.[65]

Even the demise of the Abasand plant raised many questions, to some of which Ball had certainly admitted. Given the increasing problem of war-time petroleum supplies in North America, the time had arrived when some solutions to the riddles posed by the Athabasca oil sands must be found. Thus, Abasand Oils and the Athabasca oil sands became the objects of the considerable interest of the government of Canada and private oil companies from the moment when the first producing Abasand plant was destroyed.

* * *

Three general points emerge from the history of early private enterprise in the oil sands.

First, it is clear that early commercial work followed the paths beaten by government researchers. While the theories of the Geological Survey led to the fitful attempts to drill for bitumen pools that did not exist, the practical experiments of the Research Council of Alberta redirected work toward the separation of bitumen from the oil sands that was to form the basis of all later development. The role of Karl Clark at the

Research Council was crucial to redirecting commercial experimentation along the most fruitful lines, and the Research Council disseminated accurate information (developing and promoting rather than discovering technology) that benefitted all other workers.

Second, it must be noted that Robert Fitzsimmons's International Bitumen was successful to the extent that it learned from the experiments conducted by Clark and others at the Research Council. But, Fitzsimmons's achievement in actually producing bitumen did not compensate for his many weaknesses or those of his company. Lack of capital, lack of markets and lack of effective industrial machinery all plagued International Bitumen, like frontier resource-developers everywhere. Even though Fitzsimmons was willing to pursue his goals with considerable vigour—too much perhaps for the equanimity of those who worked with him—he was unable to establish his company as more than a marginal operation. But his efforts are significant as an example of pioneer entrepreneurship. The reasons for his problems emerge from a comparison of International Bitumen and Abasand Oils.

Max Ball and his Abasand Oils also reflected the many problems of development, but Abasand seemed near to solving these technical and commercial difficulties even at the time of its demise in 1941. This approach reflects the superior technical skills of Max Ball and J.M. McClave as well as Ball's economic acumen. The Abasand venture was distinctive in that it blended American expertise and Canadian capital and resources. This was unlike the "typical" Canadian formula for economic development, summarized by economic historian Hugh Aitken as the combination of American capital with Canadian resources.[66] Only the involvement of the Canadian firm of Nesbitt, Thomson enabled Abasand to move so close to success by 1941; but only the skill of the American oil engineers encouraged Nesbitt, Thomson to invest in the oil sands. The point is that the oil sands required advanced technical knowledge and large sums of money if commercial development was to be successful. All this meant that International Bitumen, run on a shoe-string and designed by an amateur, could not possibly compare with Abasand. Large-scale rather than small-scale business and technical facilities were necessary to develop the oil sands.

Both Fitzsimmons and Ball were bucking conditions of oil supply and price which did not favour the development of an experimental project in the thirties. The conventional output from Turner Valley, for instance, provided an assured domestic supply for the Prairies. After 1936, moreover, production accelerated with the "deep-basin" discoveries there. (See Table 4.2.) Turner Valley producers themselves complained that they could not get adequate prices or market access for their output or for future exploration activity. International conditions at that time were characterized by elaborate pro-rationing and price-maintenance programmes, typified but not led by the American New Deal methods of stabilizing the oil industry. Fitzsimmons and Ball were brave indeed to promote and market a rival oil product in this environment. Fitzsimmons's view that a cabal (either business or government or both depending on his circumstances) prevented his success was a contrast to Ball's patient understanding of the myriad of factors at work in the oil industry. Both men hit upon sensible strategies in their search for markets in the northern mining camps.[67]

IV: Abasand as a Wartime Project, 1939-1945

The 1930s had been a time of private enterprise development in the oil sands despite the grim conditions the new enterprises faced. The 1940s comprised an era of renewed activity, particularly in industries like petroleum which were stimulated by World War II and the expanded government management of the economy.

During the war, a shift occurred from private to public enterprise. It involved Abasand Oils Ltd., its partners, the giant Consolidated Mining and Smelting Co. of Trail (CM&S), and the government of Canada through the Departments of Mines, Munitions and Reconstruction. The federal government, in particular, first observed, then assisted and, finally, replaced the private business developers of the sands.

The reasons for and consequences of this shift from private to public development were considerable and they must be found in the context of Canada's economic and potential situation as the nation approached and then participated in World War II.[1]

* * *

Two factors explained Ottawa's continued interest in the success of Max Ball's tiny Abasand experiment. The first was Canada's continuing dependence on oil imports for most of its needs. Canada produced only a small and unreliable portion of its requirements in the 1930s and 1940s; Canadian production comprised less than 5 percent of total consumption in the early thirties. The subsequent expansion of Turner Valley output provided a maximum of just over 20 percent of Canadian needs in the late thirties and early forties, whereupon the decline of Turner Valley and the almost complete absence of other conventional reserves led to gloomy forecasts for the future. (See Tables 4.1 and 4.2.)

Table 4.1
Crude Petroleum Production and Imports
Canada, 1936-51 (barrels)

Year	Domestic Production	Imports	Domestic Production Imports %
1936	1 500 374	35 835 264	4.2
1937	2 943 750	38 914 717	7.6
1938	6 966 084	35 101 712	19.8
1939	7 826 301	37 094 823	21.1
1940	8 590 978	42 623 484	20.2
1941	10 133 838	46 791 026	21.7
1942	10 364 796	44 120 212	23.5
1943	10 052 302	49 753 782	20.2
1944	10 099 404	57 047 826	17.7
1945	8 482 796	56 806 214	14.9
1946	7 585 555	63 407 461	12.0
1947	7 692 492	69 076 848	11.1
1948	12 286 660	77 633 690	15.8
1949	21 305 348	75 680 992	28.2
1950	29 043 788	80 124 402	36.3
1951	47 615 534	84 237 399	56.5

Source: Canada, Dominion Bureau of Statistics, "Crude Petroleum and Natural Gas Industry in Canada."

For the most part, Canada imported its oil from the United States and Venezuela. These were reasonably secure sources, unlike the Middle East oil fields, but increased consumption and the absence of new reserves contributed to Canadian fears about future supplies. Federal officials hoped that a new domestic source like the tar sands would be highly valuable.[2]

The second reason for continued federal interest was that Ottawa had maintained control over portions of the Athabasca oil sands area when it transferred control over natural resources held as "Dominion Lands" to Alberta, Saskatchewan, and

Manitoba in 1930. The view of officials in Ottawa had been that the government's previous fifty years of research into the oil sands demanded continued possession of potentially strategic lands. In addition, the Mines Branch and the National Parks Service argued that some oil sands land might yet support possible asphalt paving projects in the Rocky Mountain national parks.[3]

Table 4.2
Crude Petroleum Production,
Alberta and Canada, 1936-51 (barrels)

Year	Alberta	Canada	Alberta Canada %
1936	1 312 368	1 500 374	88
1937	2 749 085	2 943 750	93
1938	6 751 312	6 966 084	97
1939	7 576 932	7 826 301	97
1940	8 362 263	8 590 978	98
1941	9 918 577	10 133 838	98
1942	10 117 076	10 364 796	98
1943	9 601 530	10 052 302	96
1944	8 727 366	10 099 404	86
1945	7 979 786	8 482 796	94
1946	7 137 921	7 585 555	94
1947	6 770 477	7 692 492	88
1948	10 888 592	12 286 660	89
1949	20 087 418	21 305 348	94
1950	27 548 169	29 043 788	95
1951	45 915 384	47 615 534	96

Source: Canada, Dominion Bureau of Statistics, "Crude Petroleum and Natural Gas Industry in Canada."

For its part, Abasand Oils had found neither the technical nor the commercial problems of the sands easily soluble in the Depression years, although the possible markets in the mining operations of northern Alberta and Saskatchewan and the southern Territories were not illusory. CM&S had sought out both International Bitumen and Abasand Oils and had taken delivery of some small quantities of fuel oils from each company.[4] CM&S learned from these purchases that neither firm was able to assure quality or quantity of supply.

Accordingly, in 1938 and 1939, CM&S was allowed by Ottawa to undertake its own analyses of the reserves of oil sands and the separation process to release bitumen from the sands. This work was conducted, appropriately enough, by several of Karl Clark's former University of Alberta students, men like G.J. Knighton and W.D. Turnbull. Clark himself had consulted with the mining firm on this project, his first involvement in Canadian oil sands work since the Alberta Research Council had been suspended in 1933.[5] As a result of this research, CM&S vice-president S.G. Blaylock concluded in January 1939 that the firm should undertake to develop its own oil sands properties over a period of time. In the short-term, the firm would continue to deal with and through Abasand, regarding that company's operators as its experimenters. However, the mining company's officials were well aware of the great potential of the oil sands area and the fact that opportunities for company expansion were considerable.[6] As a result, the mining company continued to investigate the oil sands, at the same time dealing with Abasand and, as federal jurisdiction was maintained and government interest sharpened, keeping in contact with Mines Branch officials and their associates.

This was the situation in early 1942 when the federal government's concern with the oil situation and the general course of the war effort led to a re-examination of the oil sands as a major source of petroleum or petroleum products. Several factors led to this attention. Japan's belligerency in Asia meant that Canada faced involvement in a Pacific Ocean war as well as one in the Atlantic. Since the United States had also been drawn into the war in December 1941, Canada would be the beneficiary of its neighbour's support, but, as the Americans came to grapple with the situation in Asia and the Pacific, Canadians were forced to plan their co-operative support for American Pacific defence strategies.

The elaborate American scheme for the defence of Alaska and the Pacific was particularly important, and it involved two stages. First, the "Alaska Highway" was constructed from Dawson Creek to Fairbanks, Alaska, and airfields were built in western Canada and Alaska, to ensure that transportation links with Alaska could be maintained. Second came the development of the Norman Wells petroleum oil deposits, their transportation and refining, a plan intended to make Alaska less dependent on petroleum supplies brought by sea. The means to this was the $134 million "Canol Project", which involved the construction of a 3,000 barrels per day pipeline to a refinery in Whitehorse. The pipeline delivered crude oil from Norman Wells to Alaska from 1943 to 1945; it was dismantled in 1947.[7] The Alaska Highway and the pipeline and refinery were funded and directed predominantly by Americans, even though both involved Canadian territory and resources.

Canada was consequently under considerable pressure to make its own contributions to the continent's northern strategic defences. The potential of the oil sands for petroleum and asphalt, both valuable commodities, inevitably led to some speculation about the possible impact of the Alaska Highway or the Canol Pipeline on the Athabasca country. C.D. Howe expressed his hope that the United States Army would examine the oil sands: in this way, he thought, development would begin at last. Others, including Prime Minister Mackenzie King, took a less optimistic view, wondering whether it was proper for Americans to take over so much development in Canada's North. Indeed, British officials in Ottawa warned their Canadian counterparts that Canada should try to establish its own projects, and generally "show the flag" in its own hinterland. By early 1943, the Canadian government had begun to assert its interest in the area.[8]

Throughout 1942, Canadian business and government officials had been examining the oil sands, along with many more important resources, in light of pressures imposed by the war. Several studies by public service energy experts, by CM&S engineers, and by oilmen resulted and these enabled the government to decide whether the oil sands might be fitted into the war effort and how this should be achieved.

Early in 1942, Oil Controller George Cottrelle decided that oil sands development could substantially aid in the wartime oil needs of Canada. Surprisingly perhaps, he asserted to a skeptical C.D. Howe that "if anyone is likely to solve the problems attendant on this deposit, and which at one time seemed insurmountable, Mr. Max Ball and his associates will do it." This endorsement aside, Cottrelle acknowledged two important qualifications. First, Abasand had not yet yielded the result that even an experimental plant should have shown. Several more months of operation were needed to assess the method, its efficiency and the costs. Second, even if Abasand proved itself, it was necessary for the operation to move thirty miles to a better site in the Mildred-Ruth Lakes area where a new $10 million, 10,000-barrel plant would be constructed. Only then could Abasand contribute to the national emergency by supplying asphalt if not fuel oils for the North.[9]

Cottrelle ignored the reports of some observers of Abasand's operations, who had noted the company's tendency to operate poorly at every step from mining to refining. The skepticism of CM&S observers resulted from Abasand's long delays in delivering contracted amounts of fuel oil, as well as their experiences on field visits in 1940.[10] Early in January 1941, the Geological Survey's George Hume visited the plant and concluded that Abasand itself had not provided definitive test data; if a larger plant was built, production costs would be over $2.00 a barrel of crude "bitumen," a 40-cent loss at current prices.[11]

When the contents of the Hume and Cottrelle reports were conveyed to Abasand Oils officials they vigorously defended their company's efforts. Max Ball reported that Abasand's 600 tons per day plant had produced, admittedly only from occasional operations, 21,522 barrels of crude bitumen from 23,910 tons of sand in 1941. Of the crude extracted, 1,167 barrels were gasoline, 2,022 were diesel oil and 3,917 barrels were fuel oil. P.A. Thomson of Nesbitt, Thomson, the brokerage firm, which held both shares and loans with Abasand, reminded Mines Minister Thomas Crerar that CM&S had purchased 145,408 gallons of oil and that even more might have been sold had production begun earlier in the year.[12] Thomson did not dwell on the fact that the separation plant at Abasand had been gutted by fire in autumn 1941; neither did he report data on costs to offset Hume's estimates. Both were notable omissions.

The assertions of Abasand officials alone did not convince the government. Numerous studies were conducted in 1942 and early 1943 by the Mines Department and CM&S. Since Mines Department economist E.S. Martindale felt that George Hume's estimates were too vague, the Mines Branch Director, W.B. Timm, produced some revised estimates of his own. He presented these in a report to a joint Mines Department-National Research Council committee in April 1942.

Timm estimated the production costs at $1.31 per barrel of crude bitumen, a remarkable amount, less than Hume's figures and well within the limits of the market price. Timm's prediction led the committee to conclude that mining, separation and refining bitumen was probably technically feasible and might become economically feasible also.[13] But more work would be needed to sort out the scale of any additions to the present operation. While CM&S vice-president S.G. Blaylock proclaimed that oil could be produced at less than two cents a gallon (a very low price indeed), J.R. Donald of the Munitions Department remained skeptical because inexact information was available on the existing activities. The resolution of these contradictory assessments required further study.[14]

Another joint study was conducted, this time by a group of oil sands veterans. They gathered in May 1942 in Edmonton and proceeded to examine the Athabasca country and the Abasand site. Major G.G. Ommanney, who represented the Canadian Pacific Railway, F.V. Seibert from the Canadian National Railway, W.S. Kirkpatrick of Consolidated Mining and Smelting Company, and Earl Smith of Canadian Oils Limited (a minority shareholder in Abasand), decided that while the considerable potential of the oil sands was undoubted, no assurance of their commercial viability was yet available.[15]

This decision confirmed the skepticism and the optimism of government and industry officials. This time, however, C.D. Howe was involved in assessing what was known as the Kirkpatrick Report. The world oil situation and Canada's own dwindling reserves, plus CM&S's enthusiasm, lent weight to the Kirkpatrick Report's call for intensive, serious research. Thus, CM&S was awarded a contract by the federal Mines Branch to undertake a major assessment of Abasand's activities and to examine the oil sand reserves of the region. C.D. Howe assured Abasand Oils officials that they would be fully consulted during the assessment although the Mines Branch would administer the research funds and receive the reports first. C.D. Howe suggested that the Alberta government, which had been silent until then, would at last be brought into the matter.[16]

From July 1942 until March 1943 CM&S studied the sands deposits, mining techniques, separation processes and refining. This work involved the efforts of key CM&S researchers, including A.D. Turnbull, G.J. Knighton and W.G. Jewett, who had taken part in the work at Trail during the late 1930s. These men worked in consultation with P.V. Rosewarne of the Oil Controller's office, George Hume of the Geological Survey, and the peripatetic Sidney Ells of the Mines Branch. Karl Clark and E.H. Boomer of the University of Alberta were also consulted, as was Earl Smith of Canadian Oils. Once again an array of experienced oil sands hands had assembled, all of them drawn to a resource that still defied human engineering skills. All information was channelled to the Director of the Mines Branch, W.B. Timm, through the project co-ordinator A.D. Turnbull.[17]

From the beginning, Turnbull and his consultants sensed the potential of the oil sands, while experiencing, at the same time, the practical frustration encountered by previous endeavours. In July and August 1942, Turnbull observed that the Abasand plant had been poorly rebuilt and, he feared, almost as badly conceived. The operation suffered from faulty design and from some serious labour problems. Whereas the plant was designed to produce 600 barrels of crude per day, as yet it put out 300 or so daily—when it operated. The "second hand junk" used in building the plant and the misalliance of mining, separating and refining sections meant that Abasand was a "haywire" operation. This early conclusion was ominous in setting the tone for subsequent months of study. It also echoed a previous CM&S assessment of Fitzsimmons's primitive separation plant at Bitumount. Both the amateur, jerry-built set-up constructed by Fitzsimmons and the professional oil men's engineered operation were condemned with the same adjective: they were "haywire."[18]

Douglas Turnbull's earliest conclusion was that Abasand could not be allowed to squander the important opportunity to develop the oil sands. Turnbull and other CM&S researchers were extremely eager to assess the current prospects themselves and to plan for future development. So was the government. However, government officials had to await proof that the sands could be developed and then to deal with Abasand Oils and its persistent, articulate Canadian backers. It would not be easy to

Abasand Oils Ltd., circa 1940.   Provincial Archives of Alberta

displace Abasand Oils.[19]

Throughout the autumn of 1942 and winter of 1943 CM&S worked on core drilling, particularly on the more promising Steepbank and Mildred-Ruth Lakes areas, and subjected Abasand's operations to analysis. CM&S sent Abasand products to Universal Oil Products in Chicago—thus re-establishing the old connections between earlier oil sands pioneers, this time between Karl Clark and Sidney Blair and Blair's American employer—and awaited the results. These results reaffirmed that a refineable bitumen could be squeezed from the oil sands.[20]

However, Turnbull was appalled by Abasand's engineering and labour problems, the causes of the "haywire" operation. He found that the engineers, although they were competently led by Max Ball (when he wasn't away on other consulting jobs) and Martin Neilsen, were often unco-operative with one another as well as the operating and maintenance staff and the CM&S observers. Competition for labour, which was generated by both the Canol and the Alaska Highway projects, meant that the workforce was extremely unstable. Few men felt compelled to accept the lower than average wages and arbitrary wage-scales, let

105

alone countenance any difficulties with management. Abasand in 1942 was neither a well-run nor a contented operation.[21]

In November 1942, William Kirkpatrick of the Calgary company, Alberta Nitrogen Products, a CM&S subsidiary, reported to the parent firm's vice-president that CM&S and officials from the Oil Controller's office had agreed that Abasand was a demonstrated failure. He explained that this assessment extended from the management through the engineering to the actual operation of the Abasand plant. The facts were that no successes were found in "...digestion or vis-breaking in their refinery, that the separation plant and refinery are not in balance, and that operation is very disjointed...."[22] If the separation of bitumen from oil sands was being carried out, it was not being done very efficiently. It was also difficult even to assemble information to assess hot-water separation. For his part, Turnbull reported from the scene on 17 November that the plant had not operated since 6 November and that if it was started for the benefit of visiting "big shots from Ottawa," he would only step back and watch the "fun."[23]

Turnbull collaborated with Kirkpatrick on a summary of the potential for commercial scale production. Turnbull thought that a 10,000-barrel separation plant processing oil sand with 12 percent bitumen content could separate and recover 90 percent of the bitumen. If the plant was well engineered, he thought a barrel of crude could be produced for 86 cents; this was even less expensive than Timm's estimate. Kirkpatrick passed on more elaborately argued preliminary conclusions to S.G. Blaylock and to officials at the Mines Branch. He claimed that effective commercial separation was indeed possible, that the cost was probably under $1.00 a barrel, but that the level of war-time demand and strategic need would prove the key to any decision about actual development. One point was certain even by December 1942: Abasand could not function adequately. Still, Timm assured Kirkpatrick that a further $125,000 would be added to the $75,000 which CM&S had already received from Ottawa for its studies.[24] Conclusive information, no matter what was done with Abasand, must be provided for the government.

Max Ball continued to hold out hope for the rehabilitation of Abasand's operations and reputation, and Nesbitt, Thomson continued to insist to the federal officials that they be given the

106

opportunity to expand operation. They cited their investments of $900,000 (Abasand had of course been started in 1930) and the right of private enterprise to continue its work.[25] But the prospects were slim.

On 1 February 1942, Knighton prepared the first draft report for Turnbull that allowed CM&S decisively to advise the Mines Branch on the oil sands and on Abasand. Knighton first summarized the Abasand operation. It is a clear description of the plant's operation as designed by Ball and associates:

> *The operation consists of mining, separating, and refining. Mining consists of removal of overburden, drilling and blasting the beds of bituminous sand, digging with a power shovel and transporting the material by trucks to the Separation Plant. Separation is essentially an ore dressing operation, flotation is practiced to separate the bitumen from the silicous sand. No crushing or grinding is necessary, mixing in hot water disintegrates the bituminous sand. Heat and a limited amount of air are the only reagents required for separating and floating the bitumen. Diluent, a refinery product and similar to kerosene, is required to aid in the removal of mineral matter and water contained in the flotation froth or concentrate; this operation is done in a thickener. Tailing disposal will require a large area.[26]*

This description of Abasand's use of the "McClave" hot water separation process shows clearly how similar it was to the Clark, McClave and other methods. Most interesting, perhaps, is Knighton's emphasis that oil sands separation was a mining problem, more specifically one in ore-dressing; a mining company obviously was the appropriate developer.

The separation plant, Knighton continued, had worked for only 70 days in 1942 and had produced 9,958 barrels of crude; the refinery had worked for 82 days and had produced 712 barrels of oil, but, as the report noted, 43 percent of the production was used as fuel in the plant. It was, therefore, its own best customer. This was hardly the endorsement of a plant that might have commercial uses or which might contribute to Canada's war effort. In this regard, Knighton's report totally substantiated Turnbull's earlier dismissal of Abasand.[27]

Executives of CM&S had already offered to involve the company in the oil sands business by late November 1942. This they offered to do either alone or in co-operation with Universal Oil

Products. While the company's proposal was premature, by early spring 1943 its officials were more hopeful that their long-held wish to develop the oil sands would be fulfilled. The thirties had been an extremely active and profitable period for mining companies and CM&S had benefitted from this by expanding its activities in the north. It had no experience with the frustrations of oil sands experiments and it wanted cheaper fuel for northern operations than the oil companies provided.[28]

The federal government, however, hesitated, waiting on still another assessment of Abasand before deciding on the future of its oil sands involvement. Earl Smith was asked by Abasand's major investors, Nesbitt, Thomson and Canadian Oils Ltd., to assess the oil sands situation. Smith was a respected oilman; moreover, he had obtained experience with the oil sands as early as 1926, when he had conducted Canadian Oils' analysis for the Alberta Research Council of the potential for separating bitumen. The main reason for his entry, it might be suggested, was to try to maintain the private enterprise operation at a time when all indications were that the government would not look favourably on any extensions to the Abasand leases.[29] But if Smith took on the job in order to salvage the investment of Nesbitt, Thomson and Canadian Oils, it was conducted with the sympathy of the government. Because of this, CM&S was forced to wait. The mining men would have to defer to the petroleum experts.

Earl Smith's summary report was based on his own knowledge of the petroleum industry and on a consultant's study of commercial-scale engineering lay-outs. Smith estimated the cost of rebuilding Abasand at $268,100, a far cry from Max Ball's claim that $48,000 would make Abasand a sound plant. However, Smith pressed for a redesigned experimental operation to fully test oil sands technology. This would cost over $1 million, but the plant could produce a wide range of refined products from a 600 tons per day operation (8 percent motor fuel, 40 percent diesel fuel and 33 percent asphalt) at a cost equal to or below current market prices.[30]

Smith endorsed continued work on the oil sands. The only question remaining was to decide who would do the job. Oil Controller George Cottrelle had long felt that, in light of its experience, Abasand should be given first chance to proceed. He was aware of the problem of oil supply Canada faced and he was keen to find measures to expand Canadian output, both conventional and experimental. He assured Earl Smith that either Abasand or the federal government would continue.[31] But, because Abasand had received such a critical review from CM&S and from federal observers, it seems that Cottrelle's idea about the various options was overly optimistic and naive. Three points, one difficult to document, the others more abundantly clear, led the government to its decision in March and April 1943.

There is no indication in the public record that the CM&S proposal was ever given serious attention by either government officials or petroleum men. It is as if the mining firm, despite its years of experiments, was not taken seriously. This apparent prejudice against the miners also would appear to have included either fear of or disdain for the impressive financial strength a subsidiary of the CPR could bring to the sands. But again, the allure of the possible rewards from successful oil sands recovery seemed to induce a sense of imminent resolution to the problems of development on the part of the oilmen and the government officials. There is, however, no recorded reaction to the CM&S offer.

One other factor affecting the government was its interest in possible American involvement in oil sands development. In September 1942, C.D. Howe wrote to S.G. Blaylock that Max Ball had tried to get the United States Board of Economic Warfare (Ball later worked for this organization as his contribution to the war effort) to investigate the oil sands. Howe had been encouraged by American officials who told him that the American demand for petroleum, and possibly asphalt for the Alaska Highway, might draw on the oil sands. Howe did not share Mackenzie King's suspicion of American intentions in developing Canadian resources; rather, he and Thomas Crerar shared George Cottrelle's hope that the United States Army might begin to examine the sands. The Canadians waited through the autumn and winter of 1942-43, but nothing came of their hopes. The Canol Pipeline project, linking the Norman Wells oilfield with American military bases in Alaska began in 1942 but, by early 1943, Cottrelle reported to his chief that the American Army foresaw no involvement in oil sands work.[32]

The intractable problems of Abasand and the lack of interest displayed by the United States Army Corps of Engineers meant that Canadian initiatives were now crucial. C.D. Howe scrutinized Earl Smith's 1943 report on the situation at Abasand, and he concluded that it was "obvious" that a market for asphalt from oil sands existed at the least, and that a petroleum engineer like Smith could solve Abasand's problems. He would support continued investigation, including government expenditure, either for a reorganized Abasand or for a government-run plant.[33] But here the partiality of the government, or perhaps one might say of C.D. Howe, to allow the experts in the field to have the innings came to dominate the decision. Apart from Ells, the Mines Branch had no experts in oil sands research; neither was there enthusiasm for such work. On the other hand, Abasand Oils, or at least its principal financial backers at Nesbitt, Thomson and Canadian Oils, insisted on a continuing interest in the plant. The obvious solution was to accommodate the wishes of the government, which wanted to support development, and the concern of the private enterprisers to remain in the business.

Most importantly, the federal government had to face the fact that output from Alberta's conventional oilfields was not increasing. Turner Valley production, which had provided significant production of oil by 1936-37, had peaked in 1942 and thus domestic supply peaked. From a 1942 high of 10 million barrels (and a supply of over one-fifth of Canadian oil needs), Alberta output began a decline which the experts (if not the oilmen) agreed was going to be steady and possibly irreversible.

In this situation, the federal government hit upon a dual strategy to stimulate the oil industry. In April 1943, two federal oil industry investment agencies were created. One was Wartime Oils Ltd. This Crown Corporation was a drilling fund within C.D. Howe's Ministry of Munitions which offered loans for private oil companies to conduct development on acreages in the proven Turner Valley fields. These loans were to be repaid only by an over-riding royalty on profitable production. Canadian oilmen were only partially satisfied with Wartime Oils, for they wanted funds to explore for new oil, rather than develop a declining field. Wartime Oils provided more than $4 million for development and regained about three-quarters of this amount in royalties at the time of its dissolution in 1948.[34]

The second federal agency was a reconstituted Abasand Oils

Ltd., which saw Ball's company properties become a government enterprise. Abasand was reorganized, with its board composed of representatives from the Nesbitt, Thomson group, Canadian Oils, other shareholders and the Oil Controller's office. The Mines Branch of the Department of Mines funded and supervised operations, and the government committed $500,000 to rehabilitate the plant. Earl Smith of Canadian Oils was appointed superintendent. Abasand Oils Ltd. had finally retained both the right to reacquire the plant at fair market value and to participate in its operations. This was a complex joint venture indeed.[35]

The hybrid appeased the investors, for their interests were still recognized and their own man was in charge. More, the decision reflected the concern of responsible officials like P.V. Rosewarne of the Oil Controller's office that government financing demanded government supervision, especially given Abasand's chequered past.[36] Finally, the decision represented the Canadian initiative in the North that, for different reasons, Prime Minister King and C.D. Howe both preferred.

To CM&S, however, the decision meant the end of their aims in the Athabasca country. They were "wiped out," as William Kirkpatrick wrote. Even though Earl Smith had tried to soften the blow by an early hint to Kirkpatrick that the mining firm was to lose out to the oilmen, the company's officials were shaken. They decided to remain near the Abasands site, to continue core-drilling (perhaps as a sop, the government offered to extend a contract for further work in that direction), and they hoped they might get into the business later. At least, they would add to the $200,000 they had already received in consulting fees.[37]

The plant Earl Smith was to rebuild was seriously flawed but he did not plan on major changes. B.F. Haanel warned W.B. Timm shortly after the government's takeover that the flaws were possibly mortal. He reported on the opinion of Mines Branch and Research Council engineers, who said that the imbalance between the mining, separating and refining functions was serious enough, though probably resolvable through further engineering work. A more fundamental problem was with the diluent—so crucial to effective hot water separation. Diluent was not feeding continuously due to the very process chosen by the Abasand engineers. This meant that the separation plant worked only sporadically and that diluent would coke up during distillation after separation. Unless the original design was modified in the remodelling, Haanel prophesied, the earlier mistakes would be repeated. Rosewarne and W.H. Norrish of the Mines Department agreed, although they hoped that Smith would attack the diluent feed situation and its ramifications on the rest of the operation. But, at least for the moment, Earl Smith was in charge, and federal government mining engineers were not.[38]

Smith began to supervise the construction of new ancillary buildings as well as the reconstruction of vital separation and refinery equipment. Between May and August of 1943, however, Smith was unable to get control over a plant whose management and labour forces had long since demonstrated an incapacity to complete a project. In early August, reports of disgruntled workers and even sabotage reached Ottawa. On 31 August Peter Rosewarne found that more than three weeks earlier sand had been thrown into lubricating oils thus wrecking a number of the vehicle engines.[39] That set-back was only a symptom. By mid-September, George Cottrelle observed that even Earl Smith's superior at Canadian Oils, John Irwin, agreed that events had proceeded so slowly that Smith must be replaced. This view was sustained by the assistant Oil Controller, George Boyd Webster, following his own inspection of the Abasand plant. Webster, a mining engineer, catalogued the construction gaps for his superior in Toronto and described the plant as being nowhere near completion.[40]

Webster's analysis focussed attention on management problems. As George Cottrelle expressed it, a change of management should resolve the present difficulties and all that mattered immediately was that a plant be put together. A beleaguered Earl Smith foresaw the confrontation and resigned on 30 September 1943, handing over the half-built plant to Boyd Webster, the Oil Controller's assistant.[41] But Haanel's warning that Webster should be alert for engineering flaws remained valid, even amidst the messy process of replacing Earl Smith.

Webster, the new superintendent and vice-president, decided to reorganize the plant management and revise the work schedule in order that both would become more efficient. He also took the more significant step of reconsidering the engineering proposals for the separation plant.[42] This meant that there would be virtually a winter's delay before the reconstruction programme could result. Since much of the work done in the summer and autumn of 1943 was incomplete, a whole year of the project was sacrificed.

While plans were drawn to refabricate the separation plant, the likelihood of a larger financial obligation than the original $500,000 worried Timm. There was certainly reason for his concern because Howard Tamplin of General Engineering, the Toronto firm which had been hired to redesign the separation plant, informed Webster that an outlay of $778,000 would be necessary to complete an efficient 600 tons per day plant. The costs were now 25 percent higher, but the prospect of success made the outlay worthwhile. Timm recommended that the increased budget should be accepted and that there should be further engineering studies.[43]

The revised plant began to move toward completion only in March 1944. But, as General Engineering's Tamplin noted, the labour situation remained a weakness and cost over-runs, due to the procurement problems of a plant not given the highest priority in the war-time period, meant that progress was not nearly as great as Webster had promised.[44] Tamplin calculated that delays in receiving essential materials would delay the completion of the refinery until at least August 1944, although the separation plant was completed by the summer. Nonetheless, delays and ever-increasing costs, which sometimes exceeded $100,000 a month, meant that the situation continued to deteriorate over the summer.[45] W.B. Timm calculated with concern that monthly costs averaged $75,000 despite the fact that the refinery was not operating yet. Webster admitted in mid-August that the refinery could not be ready for six more weeks, but he claimed that "labour is the whole question." The lure of the Alaska Highway and the general boom in construction and industry in the west meant the perpetuation of Abasand's labour problems, despite the removal of the worst sources of friction, which had plagued it in years past.[46]

In response to the increasing concern of Mines Department officials, Deputy Minister of Labour Arthur MacNamara took personal charge of investigating the labour situation. The labour unrest at Abasand was caused by the company's obvious instability, the project's isolation, and the employees' relatively low wages, MacNamara explained. The plant could employ between 60 and 90 construction personnel and required some 40 operators; it was too big to depend on workers' personal loyalty, too small to generate the earnings available on mammoth, high-priority projects. The labour situation at the time was the result of the booming western Canadian labour market, not perfidy or sabotage at the plant.[47]

Timm attempted to explain the situation to the Deputy Minister of Mines, Charles Camsell, by referring to labour and materials shortages. Together they meant that the plant was still unable to conduct proper test runs. He reported, however, that the separation plant had begun to work, though spasmodically. George Hume found late in September that "…things are moving along nicely." George Cottrelle was ready to second that assessment in early October, but Webster was Cottrelle's man.[48]

Howard Tamplin was at least able to report on the separation plant, which had actually worked through much of September. The 46-man operating crew had treated 746 tons in the month, not much more than the plant's designed daily capacity, but they produced 1,065.5 barrels of crude. Still, there was 41.9 percent diluent oil in the crude recovered. Until the refinery was built, then, Abasand could not operate often as there were no large storage facilities for diluent or bitumen. As in his earlier reports of August and July, Tamplin was struck by the problem of diluent recovery—about which Haanel had warned. Nonetheless, Tamplin calculated the separation was squeezing 90 percent of the bitumen from the sands mined. In October, however—with the refinery still not running—the separation plant did not operate at all because no diluent was available.[49] This was the production of a plant which, as of 30 September 1944, had cost the federal government $1,193,728.21. The refinery was assembled on 9 December and start-up began on the sixteenth. From then on, the plant would receive serious testing. Whether it would work efficiently was still in doubt.[50]

The separation plant continued to run intermittently in December 1944 and January and February of 1945, and, as Tamplin reported, there seemed to be some improvement in the separation. Only 16.8 percent of the bitumen recovered during the December runs was actually processed diluent. As well, 16.8

percent gasoline, 16.5 percent diesel, 13.3 percent fuel oil, 34.5 percent asphalt and 21.1 percent loss was reported from the refinery.[51] This was a far better recovery rate than previously gained; at least diluent was no longer the main constituent of the bitumen. Now that the refinery seemed capable of working regularly, Abasand appeared to be on the way to becoming the sort of operation for which it had been designed.

In addition, an experimental lower-temperature separation process, using a grinder-flotation unit, was being tried out in a second separation plant. This was an adaptation of the ore-dressing techniques so familiar to many of the newer Abasand advisers. This experimental process seemed to suggest that the hot water pulping and skimming process might actually be replaceable. Abasand's hot water unit was adapted from the original Abasand operation and from Karl Clark's experimental units. Throughout its development, it had not produced favourable results. Often it provided poor quality separation and consumed high quantities of energy due to the heating of the water.[52] The new unit promised relief from these problems.

The experimental unit marked the result of the prolonged involvement of General Engineering in Abasand. Early in the new year, the ore-dressing separation process had reached the stage where "Geco" cells [after General Engineering Co.] were installed to institute the so-called cold water separation. This unit would run at room temperature, 70 degrees Fahrenheit, rather than at the 160-80 degrees Fahrenheit of the McClave hot water flotation units. Howard Tamplin went so far as to proclaim that the chief development of February 1945 was the successful operation of the cold water method. With a bitumen recovery of 94 percent to 98 percent, compared with the hot water unit's 84 percent recovery rate, the new method was promising indeed.[53]

Experimental runs using the new equipment and continued attempts to sustain operations of the older separation plant, as well as the refinery, were all continued into spring 1945. No less than fourteen of the Geco cells were ordered from the Lake Geneva Mining Co. P.D. Hamilton, President of General Engineering, recommended the abandonment of the old hot water unit altogether. Even a guarded W.B. Timm agreed that a real breakthrough might well have been achieved.[54]

The plant performed quite effectively from March to May 1944. The results indicate the scope of operations.[55]

	Sand separated (tons)	Bitumen recovered (barrels)	Recovery rate (%)
March	5051	4320	92.6
April	2953	2395	88.2
May	2203	1730	86.1

The success of the cold water unit in March in fact led to the phasing out of the original Geco cells and preparation for the construction of a third new, separate plant. While continuing to operate the hot water units, Abasand management and the Mines Branch observers awaited the new units with hope.

Meanwhile, the refinery suffered through often rough operations. The coking that the McClave process seemed to induce was perpetuated in the new plant. Again, it was hoped that the more effective Geco cells would eliminate this problem because there was higher efficiency of separation, the result of the cleaner product of the cold water cells. When these separation cells were working, they provided the refinery with sufficient crude oil to produce a blend of oils that was satisfactory. While the refinery still returned between 40 percent and 50 percent diluent (e.g. 5,210 barrels in March, 2,183 in April), it also produced gasoline, diesel fuel and asphalt. Between 40 percent and 45 percent of the refined product was "residuum" or rough fuel oil (5,040 barrels in March, 1,827 in April). In these two months, gasoline accounted for 1 to 2 percent of refined product, diesel for 2 to 5 percent of the product. But the production of useable diluent and the various fuel oils was satisfactory to all.[56]

Indeed, all awaited the new cells. On 16 June 1945, however, a welder's torch, in the sort of accident that plagues industry, ignited sand at the feeder of the old No. 1 separation plant. The result was a disastrous fire which destroyed the separation plants, the warehouse, the machine shop, the fire hall and other equipment. Fortunately, nobody was killed. P.D. Hamilton witnessed the attempt to salvage the plant but he reported that the fire caught too well and moved too quickly. Within minutes the separation plants were destroyed; the rest of the nearby structures went within the hour.[57] At the very point when some advance from the cursed fortune of Abasand appeared to have been made, this set-back wiped out the plant. Abasand seemed dogged with problems.

Oddly, initial reaction to the fire was not pessimistic. Indeed, as Webster reported to Timm, the Abasand board unanimously recommended immediate reconstruction. This time, a 1,000 tons per day plant would be built based on the new technology.[58] But there were too few results from test runs and too many problems from the past to support a precipitous decision. The board would have to ponder the future of an enterprise which had few successes and few friends.

Criticism of Abasand had plagued the plant along with its internal problems. In Alberta, the provincial government and the tiny oil industry had found most aspects of federal oil policy unacceptable. Both Wartime Oils and Abasand were viewed as ineffective means of stimulating the search for crude oil. The government of Alberta was particularly irritated by the lack of progress at Abasand.

In mid-1944 and again in early 1945, Alberta's Public Works Minister, the fiery W.A. Fallow, had severely criticized the tardy federal operation. The Conservative party press, led by the Toronto *Globe and Mail* and the *Financial Post*, had also proclaimed the failure of the federal effort on numerous occasions. Fallow and several journalists charged that Abasand's failure was deliberate.[59] Moreover, the government of Alberta, which was increasingly restive about the federal government and much else, had commenced to think about building its own oil sands experimental separation plant.

In this climate of distrust, the internal assessment which resulted from the fire led to serious doubts about the need for further development. General Engineering's P.D. Hamilton agreed that a significant breakthrough in the separation process had probably been achieved, but he warned that there should be no illusions about the commercial value of oil sands separation, at least not on the basis of data he had seen. Would the $450,000 it would take to rebuild the plant be a good investment, he later asked G.B. Webster.[60]

C.D. Howe's assessment of the situation was most important to Abasand's future. Howe, now Minister of Reconstruction, stated his initial conclusion, pending reports from Abasand to the new Minister of Mines, James Glen. Abasand, he wrote, "...should not be rebuilt at Government expense...." Sufficient experimental work was done and sufficient public investment had been made; it was up to others to carry on. There was now no pressing need to build up the oil sands, anyway. The wartime emergency was over.[61]

Glen and his department did not fully accept Howe's view. Mines Department officials still wished to sift through data on costs and operations, but that activity depended on co-operation from the tardy and prickly Abasand Board.[62] Despite the endeavours of Mines Department officials to keep open the possibility of further investment, there was little to do but survey the sorry wreck, and even the failure, of Abasand as both a private and public operation and to regret that the plant was on the brink of true usefulness when it was destroyed. Several reviews by Mines Branch officials followed, but these only verified the tentative conclusion reached in the summer of 1945 that Abasand should not rise again.[63]

Webster and Irwin tried to estimate the cost of separation and refinery operations with a view to refinancing Abasand by Ottawa. They were working on the basis of very incomplete data: so much information had been lost in the fire that even a proper analysis of operations was impossible. Webster noted that the plant had cost $65,000 per month from April 1943. At the production level Abasand had reached, he determined a $2.197 per barrel cost of crude bitumen. Since the Canadian price of crude oil in September 1945 was much less (somewhere above $1.00 per barrel, depending on the location of the market), this was not a good sign. Webster's extrapolation of the cost of producing in a 1,000 barrel a day separation plant was $2.31 a barrel; he did not consider the cost of shipping the oil from the Athabasca country to any potential market. Finally, when the final official report of the Abasand board was presented (it had been demanded by Timm so that he could give a definitive response to the board's request for further funding), it determined that crude bitumen would cost $2.51 per barrel to produce. The report asserted that this was not far off an economic price, given the complicated subsidies on the price of oil in western Canada and the high costs of transportation.[64]

Like the oil sands, this argument did not wash too well. Timm had to report to his supervisors that neither hot nor cold water separation promised to yield a commercial product. He himself extrapolated a cost of $2.226 a barrel on the $65,000 monthly costs and 29,600 barrels of net crude produced. When Webster and Irwin attempted to argue in favour of the economic feasibility of oil sands production, they convinced no one at the Department of Mines.[65]

The total cost of the Abasand operation added up to over $1.9 million. With this in mind, would existing circumstances justify further expenditures? Mines Minister Glen himself seemed unconvinced as early as October 1945. If a final decision did await much business and bureaucratic sifting and lobbying, it could not be expected to cinch a case for further investment. Yet for almost a year, arguments were reiterated, data were repeatedly cited, and hopes were raised. Since C.D. Howe had spoken, the Mines Branch was not going to act hastily. In the summer of 1946, James Glen began to soften against the lobbying of Abasand and Nesbitt, Thomson. He floated a hesitant suggestion to revive public investment in Abasand. Howe then stated his firm opposition. And that was that. By June 1946 the federal government admitted it would not re-enter the oil sands industry. By then, Alberta's new plant at Bitumount was under construction.[66]

It should be recalled that the federal government had only a limited and haphazard record in managing petroleum resources, as is suggested by the squandering of Turner Valley oil and gas. It is instructive that a recent study of Canada's national fuels policy sees fit (and fairly so) to ignore oil and gas as a subject for national resource policy discussion until the pipeline debates of the late 1940s and 1950s. While it is clear that further study of early petroleum policy is almost certainly needed (the continuing Alberta hostility to Ottawa's interventions from the 1950s through the 1980s has its beginning in the period covered by this book), the fact is that no co-ordinated federal policy existed. The double programme of Wartime Oils and Abasand Oils established in April 1943 was a noteworthy enough policy development.[67]

Abasand's remaining equipment was not disposed of until 1948. Fire insurance monies had been divided among the various interested parties in late 1946, after much squabbling, and Abasand was returned to its private owners. They had not paid one cent for the federal government's investment, as this had gone up in the flames of June 1945. Only in the 1950s would Abasand become active again, and then to align itself with the Sun Oil Company in a larger scheme for the oil sands, one which would result in the Suncor plant miles north of the Abasand operation.[68]

* * *

In many ways the government's Abasand plant was as much a failure as the private operation. Despite the nearly $2 million invested by the federal government (and the nearly $1 million spent by Abasand Oils Ltd. previously), the plant could not sustain the feasibility of oil sands development. To conclude that public enterprise could not succeed would be tempting, yet this is too easy.

The fact that the federal investment went so badly must be seen as the result of the intractability of the problems addressed and of the novelty of an initial attempt at co-ordinating an oil policy for Canada.

Concerning the technology of development during the war, it can be suggested that reliance on the discredited efforts of the petroleum engineers should more quickly have been abandoned for the fresher theories of the mining men. However, in light of subsequent developments, it cannot be claimed that a magical solution to the oil sands mystery was ignored. Since it has been shown subsequently that the hot water process was the most practical, the problems at Abasand must be seen as intrinsic to the improvement of technology and the exploitation of the resource. Problems were the result of neither perfidy nor mystery, as too many local historians have inferred. Indeed, the experience of the federal government at Abasand was important to subsequent development and was thus extremely valuable. It showed that even the most competent of officials and experts could not easily resolve the oil sands problems. Moreover, the $1.9 million of federal money was hardly a significant waste to a government which had spent over $5 *billion* in 1945 alone. Since most of that money was spent wisely, it is not well founded to criticize government enterprise as being the cause of the Abasand failure.[69]

The federal takeover in 1943, and frustration up to 1945, stimulated the demand for even more federal money on the part of Abasand Oils Ltd. It therefore set the pattern for private enterprise to seek government help in covering the costs of oil sands development. Moreover, the federal effort directly stimulated Alberta to reassert provincial control and to begin the Bitumount project. In this way, Ottawa's failure served to prompt the type of further development that had interested the federal and provincial governments in the first place. And, it certainly reminded Alberta to keep closer control over its own natural resources. In sum, the stimulation of both private business and the provincial government lies behind the advances in oil sands technological and commercial knowledge which occurred in the late 1940s and early 1950s. This knowledge, in turn, reached the point where illusions about the oil sands were at last abandoned. The Abasand plant was a modified failure only for it achieved innovations in the development of the oil sands.

The Refinery at Bitumount, 1947-48.   Provincial Archives of Alberta

V: Bitumount The Alberta Government and the Oil Sands, 1945-1951

The problems at Abasand confuted any hopes for easy development of the oil sands and the federal government was discredited by the role it had played. Nonetheless, evidence gained from Abasand's actual operation offered some limited grounds to hope that separation was practical and that processing might be economically feasible. The difficulties of the federal government's operation, a plant that worked only by fits and starts through 1944 and 1945, occurred at a peculiar time for Alberta's Social Credit government.

The government of Alberta had as one of its goals the development of the province's resources. This, coupled with its dissatisfaction with the federal government, led it to re-enter oil sands development. By early 1945 it had committed itself to build its own experimental separation plant in order to sustain the private enterprise which, both levels of government agreed, must be behind any effective commercial operation. The post-1945 resurgence of major provincial intervention and the construction of Bitumount proved to be extremely trying to the provincial government, given its goal of securing private development of the oil sands. The intervention was also ironic given the government's ideological motives.

Bitumount was at least a modified technical success, but its equivocal results were combined with major changes in the

Alberta oil industry after 1947. The result was that plans for oil sands development were stalled for over a decade after its completion.[1]

* * *

The Alberta government had not been comfortable with the federal government's development of a portion of its provincial natural resource. This discomfort took two forms. Political leaders like premiers William Aberhart and Ernest Manning (Manning became Premier when Aberhart died in May 1943) and Public Works Minister W.A. Fallow based their objections on the view that Ottawa's progress hindered private initiative in the oil sands. In 1943, Alberta had agreed only to allow Ottawa time for experiments at Abasand. But the province grew impatient at the continued problems at the Horse River Plant.[2]

The other source of provincial hostility lay in a more general conflict between Alberta and the Dominion government. After 1937, Alberta had attempted to implement some major portions of the Social Credit doctrine, a course which had involved financial legislation that did not conform easily to the division of powers laid out in the British North America Act or with prevailing ideas about such policy. Alberta was involved in a long conflict with the federal government in the courts over its debt adjustment legislation. In the end, the Social Credit government was unable to overcome the constitutional limits on certain of its actions. The result was that the Social Credit party, if not the government, was extremely hostile to the federal government. Moreover, the interventionist orientation of Prime Minister King's war-time administration and its acceptance of systematic social welfare programmes as part of its post-war reconstruction scheme antagonized Manning's conservative and autonomy-oriented government as well as the doctrinaire Social Credit Board (at least until it was disbanded by Premier Manning in 1948). The point was that Ottawa's interventions into Alberta's affairs proved generally unacceptable to the provincial government and for groups influential in the province.[3]

One result of these political and ideological conflicts was increasing dissatisfaction over Abasand which, although a minor business, was viewed with annoyance by the provincial government, especially as the company became increasingly

124

peripheral to the war effort and as federal-provincial tensions increased.

The plans of Lloyd Champion, the proprietor of International Bitumen from late 1942, gave the province the excuse it needed to argue that commercial development was being inhibited by the Abasand plant. Champion, the Montréal trust company executive who replaced Robert Fitzsimmons and renamed Fitzsimmons's firm Oil Sands Ltd., attempted somewhat unsuccessfully to refinance and to recommence operations at Bitumount from 1943 to 1944. He had sought support for his project by approaching the American government with proposals to sell bitumen for road oil or fuel oil. Here he collided with the Canadian government project. He then turned to Alberta, seeking support for his marketing strategy. To the Alberta government, Champion represented the type of private developer who would solve all the problems of the oil sands and realize their great potential. Understandably, the Alberta government was eager to see Champion succeed. When Champion found that he was unable to raise the private capital or sign the government supply contracts necessary to begin large-scale development, he sought the co-operation of the provincial government either in a joint project or in refinancing his firm.[4]

Champion's proposal came at the time when Alberta was eager to carry out Premier Manning's January 1944 argument to the Minister of Reconstruction, C.D. Howe, that Alberta could no longer wait for Abasand to prove the value of the oil sands. Although Premier Manning proposed to open the oil sands to private developers, only Lloyd Champion had proposed a co-operative venture with the province by autumn 1944. It was a measure of the province's desire to see oil sands development occur, and a measure of its annoyance at the federal government, that it entertained Champion's proposal, which went against its small-government, free-enterprise ideology.

The government nonetheless considered the Champion suggestion with some caution. The ideological conflict within an avowedly free-enterprise Social Credit, then undergoing its right-wing's own anti-socialist campaign, must have been somewhat difficult to reconcile. In its attempts to evaluate the Champion proposal, the government called on the long-time oil sands expert, Professor Karl Clark. Replying to questions from Lands and Mines Minister Nathan Eldon Tanner, an able and influential member of the government, Clark supported the general programme of an experimental separation plant to be constructed on the Bitumount site as a joint public-private enterprise. The technical problems of oil sands separation were readily soluble, he argued, but they had to be worked out before commercial-scale construction could begin and before private business would be convinced of the feasibility of commercial production. In his conclusions Clark concurred with another of the government's advisers, oil sands veteran Martin Neilsen. A former employee at the Abasand plant, Neilsen felt that the engineering problems involved in the oil sands required further experimental work. He and Clark thought that such an experimental plant should prove that oil sands separation was economic. As Clark argued, "if the cost is not established soon, the resource may lie idle, needlessly, for years before some outside party ... appears."[5] Clark, at least, was concerned about any delay, probably because he knew that exploration for conventional oil was underway in the province.

Champion's formal proposal, put forward 13 October 1944, called for a government-business partnership similar, ironically enough, to the Abasand agreement. Champion sought government funding, either through a loan guarantee of $250,000 or a direct loan of $125,000 with $25,000 increments of working capital as needed. In return, a joint board consisting of Oil Sands Limited staff, Research Council of Alberta scientists, and cabinet officials, would form a board to supervise construction and operation of an experimental-scale plant. Oil Sands Limited would retain all commercial rights while agreeing to pay back the loans it received.[6]

This offer was not acceptable to the government. Premier Manning wrote to Champion that he agreed to a cash-advance method of funding a new plant at Bitumount, but the government wanted a direct partnership with Oil Sands Limited. The final agreement reflected these revisions and on 6 December 1944 the joint venture was sealed. A Board of Trustees was appointed, composed of W.A. Fallow and N.E. Tanner from the provincial cabinet and Lloyd Champion from Oil Sands Limited. The Board of Trustees supervised the operation and the

Provincial Marketing Board handled financing and purchasing of materials for the plant. This, even more than Champion's proposal, was a mirror image of the Abasand structure after the federal government became involved. Given the federal-provincial political conflicts of this period, it is even more ironic that $250,000 of funding was to come from a special "Post-War Reconstruction Fund" that Ottawa provided each of the provinces.[7] Significantly, the estimated cost of building a good experimental plant did not seem to reflect the uncontrolled and accelerating expenses that Abasand had incurred. It remained to be seen whether the province could build a plant either more efficiently or more cheaply than the Dominion.

The ease and optimism with which the two partners entered into their project did not suggest that they had learned much from the unfortunate experiences at Abasand, although Abasand was moving toward the solution of its technical problems by late 1944, and there was some basis for hope. Since the problem of joint government-business direction had delayed the Abasand project, there was also some reason for more caution. As astute an observer as Karl Clark had explained that the Abasand situation was the result of "government ineptitude," rather than federal sabotage, which W.A. Fallow had suggested was at fault.[8] Yet Clark blithely assumed that the Alberta joint development project would not face problems of government ineptitude, perhaps because he was confident that Alberta would use oil sands experts for advice on building its plant.

Lloyd Champion, indeed, was soon asking his fellow Trustee, N.E. Tanner, to help bring about closer co-operation between the Board's advisers and Oil Sands Limited's technical staff. The project was not getting underway quickly and greater co-operation was essential.[9] Very little was done, however, in the project's first year. Excluded from formal participation, the Research Council, like the Mines Branch at Abasand, was called on to assess the designs for the new plant. A Tulsa, Oklahoma firm, Born Engineering, which had no previous oil sands experience, had been hired to design the plant and took a long time to produce plans. Karl Clark at least found much to criticize in the blueprints drawn up in 1946.[10]

About this time, Sidney Blair, who had been active in oil sands research with Karl Clark, wrote from Toronto to his old friend enquiring about the state of the project. Blair, who had spent some years with Universal Oils in Chicago, followed by wartime work for Trinidad Leaseholds, wanted to re-establish himself in Canada. Should he continue to demonstrate an interest in northern Alberta resource development, he asked.[11]

In response, Clark outlined both the history and current status of the Bitumount project. He began by explaining that the "real motive" behind the plant's construction was the provincial government's intense loathing of Ottawa. This dislike had led the province to try to prove that Ottawa was not committed to the success of the oil sands. The result was the province's commitment to build its own trial separation plant. When Champion made his proposal, Alberta was eager to proceed with a partner from a private firm.[12]

Clark characterized Champion as "the business brain and the boy who could roll all the politicians and others that needed rolling," but commented that he had since "proved a washout so far as a good businessman is concerned." This meant that the project was not going ahead with the speed that the government had counted on. Clark noted that the engineer in charge, Martin Neilsen, an experienced oil and oil sands engineer, was less able at dealing with men than with designs. Again, this acted to the detriment of the project. Clark also remarked that the Research Council and he himself were only informally consulted, although that had not meant that their advice was not counted on. He noted that the Research Council would be given the "run of the plant when it was erected and in operation...." Finally, he observed that the plant was a long way from completion but that the original $250,000 was long since spent. Indeed, the project had cost $428,000 by the summer of 1946 and it was still incomplete.[13] Clark gave no indication that he thought that the plant would not be completed, but he gave every indication that the organization running the project was not working very well.

Karl Clark's account suggested that Bitumount was beginning to take on some of the traits of Abasand. It was not until the late summer of 1947 that the plant actually neared completion, although it was still a long way from operating. By then, Clark had become more than hopeful about the plant, even though he observed then that it had cost over $500,000 (twice its original

128

Manager's cabin (right) and workers' bunkhouses at Bitumount, 1947.
Provincial Archives of Alberta

budget) and would require another $250,000 before operations could begin.[14] Clark described the nearly-complete plant as "dandy fine," claiming to Sidney Blair that it was "streets better than the federal government's effort at Horse River." He continued his description:

> We have a better plant, better equipment, all new, good instrumentation all on top of better tar sand to handle and right on the river 60 miles nearer the market and the cost will not exceed the $750,000 figure. . . .[15]

Clark's enthusiasm was such that even he smuggled in hopes for the commercial viability of the Bitumount plant and of the separation method that he had, a year earlier, decided was not the key to commercial development. (In 1946 he had argued that only an *in situ* recovery method could tap sufficient reserves cheaply enough to make the oil sands an economic resource.) It is a comment on the El Dorado lure of the Athabasca deposits and on the excellent facilities that were then being built at Bitumount that Clark could be so swayed by their prospects in the summer of 1947.

However, the partnership administering Bitumount was not faring as well as its construction site. In late 1947, Lloyd Champion thanked Premier Manning for the "reinstatement" of Oil Sands Limited's leases of 3,300 acres at Bitumount and for 6,700 acres of leases in the Mildred and Ruth Lakes area. These developments, Champion concluded, gave Oil Sands Limited a very good chance of refinancing itself. It would now be attractive to the "big financial interests" in the east.[16] This "reorganization"—and further documentation, as opposed to press gossip, on this point is scarce—suggests that Oil Sands Limited was finding itself unable to sustain its part of the Bitumount project.

129

The firm defaulted later in 1948, and the Oil Sands Project became a wholly provincial government undertaking on 20 November. Temporarily at least, Oil Sands Limited was removed as a private developer of the oil sands. The firm was not totally out of the industry, however, for it did not forfeit its Mildred and Ruth Lakes leases, although it did surrender the ones at Bitumount.[17]

As for the provincial government, it alone was to continue the project and to accept responsibility for the outcome of the experiments at Bitumount. N.E. Tanner, along with D.B. Macmillan, Fallow's successor as Public Works Minister, and Industrial Development and Labour Minister J.L. Robinson, then comprised the cabinet's Trustees. George Clash of the Provincial Marketing Board, the purchasing agency for the project, and the new plant superintendent, W.E. Adkins, the former employee of Fitzsimmons and of Born Engineering, rounded out the reconstituted Board of Trustees.

Before this shake-up occurred, however, Bitumount had faced a more immediate threat. On 24 May 1948, the machine shop and warehouse had been destroyed by fire. While the separation plant and refinery, adjacent to the shops, had not been affected, both the loss of spare parts and the narrow escape from a general conflagration were reminders that the project could easily suffer the misfortunes of Abasand. In any event, the plant would not be running until August at the earliest; virtually the whole 1948 season would be lost.[18]

Bitumount did start up in late summer 1948. The usual start-up problems of any industrial plant and the lateness of the season restricted tests to a very few hours of operation. The summary of 1948 operations was presented through the Board of Trustees by the Project engineering staff to the Provincial Legislature as a Sessional Paper tabled in spring 1949. It was the first systematic report on the plant since its inception. The report began with a defence of the need for the project. Canada was still dependent on foreign oil supplies and domestic demand was still growing. Because of this, the oil sands project was still a crucial source of domestic oil. Thus, the pilot plant at Bitumount was a necessary means of demonstrating to private investors the possibilities of the Athabasca sands.[19]

The report did not summarize the plant's limited hours of operation; perhaps they were too few to be admitted. Instead, the study acknowledged several problems that the autumn tests had revealed. Water supply was a potentially serious problem; due to low river levels in the autumn, the separation plant was starved for sufficient water to support long-term operating. More important were difficulties with the feeding of sand into the separation plant and also the removal of tailings from the flotation cells. This problem had drawn Karl Clark into controversy with plant designers, as his suspicions about the designs for sand feeding and removal were proven correct when constant break-downs occurred. The major problem, however, lay with the flotation cell, the tank where the slurry of oil sands mixed with water and diluent enabled bitumen to be skimmed off, leaving sand tailings. There was an apparent "inability of the skimming conveyor to remove the oil as fast as it was produced," which demanded a slower feed-rate, in addition to revisions in the skimmers. The report indicated, though, that adjustments to both the feeding of oil sands and the water and diluent mix as well as to the skimmers were each within the capacities of the equipment. The report proved this capacity for adjustment by noting that production had increased from the test run average of 172 gallons per hour to 516 gallons per hour during final tests, the latter of which worked out to a 355 barrels per day capacity, a fair bit short of the designed capacity of 500 barrels per day.[20]

The study also reported on the costs of the project. It noted the relatively high transportation charges levied against the more than 500 tons of goods transported in each of the years from 1946 to 1948. In 1948 freight charges had declined somewhat from earlier years to $12.06 per ton. More than 450 passengers were carried to the site during the three years of work; the airstrip completed during 1948 would relieve that problem. Labour costs had also increased during the three years of construction, from 83 cents per hour on average in 1946 to $1.10 per hour on average in 1948. The total payroll had been over $75,000 in 1946, $72,000 in 1947 and $71,000 in 1948, at which time 48 men were retained as operating staff. The plant buildings had taken two-thirds of the $750,000 cost of the project, with materials for each of the separation plant, refinery and power house costing $124,570, $72,300 and $68,025 respectively.[21]

While the results of the 1948 operations pointed out the problems to be addressed, they also showed that Bitumount was likely to run effectively. In this regard it was every bit as "dandy" as Karl Clark had described it to Sidney Blair. For its part, the provincial government must have been relieved at the message of the 1948 report. At least Bitumount was not to be the failure that Abasand had ended up.[22]

The 1949 season was to be the year in which full-fledged tests would be conducted and the oil sands proven to be a resource capable of development. In the event, the plant withstood tests well, but the results were not quite as unequivocal as Bitumount's backers doubtless had hoped.

It should be noted that once the province took over the Bitumount project, Trustees' meetings were recorded and minutes of the meetings have survived. Only for the 1949 season, then, does such detailed evidence about Bitumount's management and production remain available.[23]

Early in 1949, the Board of Trustees began to plan for the season which would prove the viability of oil sands development. Various aspects of working conditions and wages were revised, including a 10 percent isolation bonus added to wage rates, since these were frozen at 1948 levels. As well, the Board announced its intention to see a refund on federal sales tax charges during the project's history. The federal government later agreed to this exemption and returned $8,259.23 to the province. Plant superintendent Adkins was emphatic on the need to treat labour fairly so that some of the trained personnel from 1948 would return and so that the labour turn-over would be minimal.[24]

Later in the spring, Adkins reported to Board Chairman D.B. Macmillan that current funding of $15,000 per month for the fiscal year 1948 was not sufficient and that it should be raised to about $18,000 per month in 1949. This escalating expense may have been worrisome, but it was far less than the $70,000-80,000 per month that Abasand cost during its final frenzied months of construction. Moreover, the escalation was hardly out of line in a period in which inflation was running over 10 percent per year.[25]

In April and May, the plant had not operated. This was due to training and retraining its crew, the result of the winter lay-off which Adkins had not favoured; insufficient water supply because of low levels in the Athabasca River had also been a problem. Indeed, the question was raised by Adkins whether a new, costly water supply was needed. The problem was that the plant was located on the east bank of the Athabasca, a very shallow channel which experienced extremely low flow volume in spring and autumn. Fortunately the river's volume increased temporarily during June 1949.[26]

By early June, Adkins was reporting on the resumption of trial runs and on 16 June he reported that the plant had simultaneously carried out mining, separation and refining operations. Earlier, on 4 June, the separation plant produced effectively at the rate of 660 barrels per day. It delivered a bitumen product with 12 percent water content to the dehydration unit. Through 16 to 19 June, 300 tons of sand were processed each 24 hours and 215 barrels of crude produced from the refinery. At that point, the plant was shut down, because tailings were not being dammed effectively and there was spillage into the river—even in 1948,

Bitumount Plant, 1948.   Provincial Archives of Alberta

133

such emission of pollutants was considered unacceptable. In sum, Adkins concluded on 21 June, the "chief difficulty" during the first weeks of work had been the inexperience of the crew rather than the breakdowns and technical flaws. But he conceded that the present crew was co-operative and teachable.[27]

When operations resumed late in June, the separation plant ran at an acceptable level, but the water supply again proved insufficient for all-out production. Adkins reported that the water level and the more serious problem of the shortage of diluent were hampering the refining operation. The diluent shortage was due to problems with the recovery system in the refinery. Adkins asked Sidney Born, the head of the Tulsa design-engineering firm, to comment on this. Apparently the situation was remedied, for no further problems with the diluent supply were noted.[28]

The plant operated routinely through most of the summer and early autumn of 1949, although two problems disturbed the equilibrium of both plant operations and the management team. One of these was the water supply. Adkins continued to monitor river volume at Bitumount. He was convinced that the primitive intake system would lead to the curtailment of plant operation in the autumn, if not sooner. In July, he strenuously urged the Board to consider the outlay of the $40,000-$50,000 needed to build a longer intake pipe and a larger capacity pump system, otherwise, the plant would continue to be less than effective. By the end of August, Adkins warned that the problem would result in Bitumount's closure at the end of September. What this meant, he explained, was the curtailment of an operation whose equipment and crew were now working in splendid alignment. If the plant closed, could the now-seasoned crew be reassembled? Would Bitumount be able to remain a model test plant? Adkins clearly hoped that it was intended to continue operation beyond 1949, but he did not seem to sway his fellow Trustees.[29]

Another source of contention was the treatment of Adkins's crew. He angrily told the Chairman of the Trustees' Board that George Clash had tried to cancel the crew's 2 percent holiday bonus pay. He also complained about the problem of tardiness in paying employees' wages. Because Clash insisted on having cheque-signing authority, after pay sheets were mailed to Edmonton and

Machinery in the Bitumount Separation Plant, 1947-48.

Provincial Archives of Alberta

cheques mailed back, pay was often two or three weeks late. There was, therefore, a three or four-week delay in getting paycheques back to banks in the south. The resulting dissatisfaction in a crew which already felt its insecurity of tenure, threatened to upset the men still further. Adkins's concern for his employees' well-being and his plant's future operations led him to recommend to N.E. Tanner that 21 of the 48 workmen be kept on over the winter of 1949-50. Their wages, which varied from $1.00 to $1.25 per hour for separation plant and refinery operators and $1.25 for shift engineers to $1.45 for the welders, had not increased greatly from those at Abasand, despite the post-war inflation. Adkins was right to be concerned about keeping his crew.[30]

The importance of an experienced crew to both the immediate and the long-range operation of the plant was exemplified by a potentially-disastrous and nonetheless serious accident which disturbed the plant on 29 June 1949. An explosion seriously burned a plant welder while he was working in the separation plant, because his careless assistant had been cleaning equipment with gasoline prior to repair work. The welder was rushed to Fort McMurray's hospital by airplane and recovered from his serious burns, but the accident pointed out the potential danger to human life and damage to the plant which might be caused by inexperienced workmen. The plant was in fact shut down until 11 August. Safety precautions and staff training remained uppermost in the minds of the operating staff at Bitumount. No doubt the experience at Abasand, where accidents as well as job dissatisfaction had so regularly harmed the project, must have been in Adkins's mind during his struggles over wages and conditions. Bitumount, indeed, had an uneven safety record. There had been three lost-time accidents in 1946, four in 1947 (the years of construction) and one during 1948 and two in 1949. Given the relatively small staff, this was not a sterling record; because of Bitumount's isolation it was doubly worrisome. An emphasis on safety was a continuous and necessary prudence given the dangers to the project and its employees.[31]

Adkins also continued to press for the rectification of the water problem and, thus, the long term capacity of the separation plant. His pleas to Tanner and Macmillan apparently received little encouragement.[32] His reports on production, however, were highly satisfactory and more obviously received with pleasure. On 13 September, for instance, he reported that 318 barrels of crude were produced in 24 hours and that the plant could regularly beat its rating of 500 tons per day of oil sands processing. Karl Clark confirmed this favourable impression when he referred to Bitumount in his private correspondence as a test plant that has "almost ceased to be interesting," because it was working so well.[33]

The season of 1949 also saw a visit of forty members of the provincial legislature to the site. Between 24 and 27 August, the little camp entertained this stream of undoubtedly curious politicians in order to show them where and how the best part of $1 million had been spent. The trip imposed a myriad of logistical problems which had pre-occupied George Clash during the summer and which had annoyed a production-oriented Adkins. The visit was a success, but it almost proved disastrous. On 26 August, a small explosion occurred in the separation plant. Fortunately the heater which exploded was repaired on the spot and the MLAs were able to witness the plant actually in operation. The $1,539.75 that the trip cost the project, then, was money well spent, although it might have enabled the visitors to watch the plant's destruction. If the explosion was a bit of ill-luck, it also demonstrated that Bitumount was not rent by the misfortune Abasand had been.[34]

On 30 September, operations at Bitumount ended because the river's water levels had fallen. The plant's record and future were then to be evaluated. The plant's three components each reported their operations; mining activities had gone on for 265.25 hours and the operations had yielded 14,866.5 tons of oil sands; the separation plant had produced 408,000 gallons of diluted crude bitumen or, subtracting diluent, 274,624 gallons of bitumen; some 399,945 gallons of diluted crude were charged in the refinery; 255,279 gallons of fuel oil and 139,432 gallons of diluent were recovered. While 11,831 gallons of fuel were on hand after the 1948 season, over 105,444 gallons were left at the close of 1949. The bulk of the work in each segment of the plant had occurred in September. Moreover, the bulk of efficient production had occurred then too. Table 5.1 illustrates the efficiency that Adkins and Clark noted on the September 1949 activities.[35]

The plant's September production was most satisfactory and that one month gave an obvious indication that Bitumount was a good plant and the process of hot-water separation was a good process. It was, after all, capable of considerably greater work-horse activity than the Abasand operation, and capable of greater output than its own limited water supply allowed. Its capacity to sustain production in the range of 400 gallons per hour from late August to 30 September contrasts with the intermittent production levels at Abasand. Karl Clark and the Research Council provided more detailed studies of flow rates and heat balances that would be used later in studies of the economics of the separation plant and also estimated production costs. Clark thought that the production of crude bitumen was about $2.55 a barrel with refining adding $.74 a barrel. Such a total cost—$3.29 per barrel was only approaching the costs of conventional oil ($3.22 was the average well-head price in 1948 in Alberta) but the small scale of Bitumount and the

Table 5.1
Bitumount Plant, 1949 Operations

Component of Plant :	June	July	August	Sept
Mining tonnage hauled:	2003½	3098	2945	6820
Separation operating hours:	149	212	139	320
net crude produced: (gals.)	43,048	53,500	55,468	122,608
production rate: (gal./hr.)	288.8	252	400	383
recovery: (bbl/ton)	n.a.	.495	.54	.514
Refinery operating hours:	62½	133	121½	263½
fuel oil: gals.	40,051	54,822	50,097	110,249
diluent: gals.	13,687	35,461	31,402	58,882
refining loss: %	1.32	.865	2.47	.42

Source: Compiled from "Summary of Plant Operations, 1949" (by W.E. Adkins) in Provincial Archives of Alberta, Public Works Department Ministers' Files, 1949, 1.

addition of transportation costs meant that any estimate was highly unreliable in so far as determining the commercial potential of the product. Clark also estimated that Bitumount recovered about 90 percent of the oil in the very rich Bitumount deposit (a deposit which contained up to 17 percent bitumen). In comparison, at Abasand or International Bitumen, about 75 percent recovery was considered good and even the Research Council laboratory had not done better than 80-85 percent recovery. By this test, then, Bitumount was indeed impressive.[36]

Finally, the total expenditures on Bitumount to 30 November 1949 amounted to $725,000. The funding had come from the post-war reconstruction fund with additional money from special warrants under Alberta Legislative Appropriation 729. Karl Clark's 1947 estimate of the plant's cost was fairly accurate. The plant had cost about one-third of the amount expended on the Abasand plant. By the end of its period of operations, in 1955, the plant had cost the Alberta government a total of $1,028,820.37; its total cost, therefore, was half that of Abasand.[37]

The issue facing the Board of Trustees and the provincial government was the future of the plant, a problem about which the Board itself was not clear. The Trustees had not been willing to fund work on the water supply system and only six men had been kept on during the winter of 1949-50. In fact, the Board was left wondering whether and how the year's work should be analyzed. There were two or three alternatives to be weighed.[38]

W.E. Adkins argued that little could be gained from further seasonal operation. On the contrary, he continued, "further studies of any appreciable scope will involve the expenditure of considerable sums for expanded plant facilities." Since the costs of developing all-weather facilities, an adequate water intake and proper storage facilities were so high, the best course appeared to be to make Bitumount available to private enterprise, which could then develop it as a research and experimental facility. Adkins noted that Universal Oils of Chicago was again interested in the oil sands and suggested that it might be pleased to take over the plant. These telling points were made to a government which was not pleased to be spending money on a venture in which it wanted private business to act, but Adkins had not specified how or for what amount Bitumount could be transferred to the private sector.[39]

An alternative plan was more carefully worked out by Karl Clark and Sidney Blair. While Adkins was not present at the Board's deliberations of 26 November 1949 (he was away in Ottawa touring the federal government's research facilities and work on oil sands), Clark and Sidney Blair as visitors laid out an alternative. They argued that Bitumount had fulfilled its main objective, which was to prove the technical feasibility of the oil sands separation process. Accordingly, Sidney Blair offered to conduct the detailed evaluation of the data from Bitumount. That evaluation, he suggested, would resolve the question of the commercial feasibility of the oil sands.[40]

It is clear that this proposal had the merit, at least, of implying a definite course that would lead to development. Blair promised to provide the answers to questions about the economic and technical practicability of oil sands development. Moreover, as an experienced and reliable international petroleum consultant, Blair could provide a convincing programme for the possible commercial development which might follow if his report foresaw commercial potential. Blair then submitted a detailed proposal to study the economic and technical aspects of mining, separation, refining and transportation of the oil from oil sands to D.B. Macmillan. Blair asked for $7,500, including expenses, for a one-year project, plus $17,000 for third-party expenses. His offer was accepted on 5 December.[41]

W.E. Adkins returned from Ottawa too late to propose a new option. After his visit to the National Research Council work on "fluidized solids" separation and to the Mines Branch work on "cold-water" separation, Adkins decided that there was merit in continued experimental work at Bitumount. He suggested that the plant at Bitumount could become the site for continued government-sponsored research, including joint federal-provincial experiments.[42] This *volte-face* on Adkins's part was too late, too ill-proposed and probably too conciliatory toward Ottawa to be accepted by the Alberta government. (The government had followed up Adkins's initial suggestion about Universal Oils by soliciting the Chicago firm's response. The company did not respond.)[43]

One fact affecting any decision about Bitumount was that Alberta had become the scene for considerable discoveries of conventional oil reserves. In 1947 and 1948, Leduc and Redwater fields were found to contain over 200 million barrels and 700 million barrels of oil respectively, more in total than the 800 million barrels estimated by adherents of the oil sands mining project. The result of these discoveries was that the oil industry was concerned with conventional endeavours rather than experimental ones.[44] The government was not overwhelmed by inquiries concerning the oil sands. Thus, Blair's concrete proposal was the one which actually promised the only effective move toward commercial development, and that remained the goal of the province. Alberta and the oil industry would wait exactly twelve months while Blair prepared his study of the oil sands based on Bitumount's operations.[45]

The importance of Bitumount to the future of the oil sands as of

1949 can be seen from the fact that it provided the crucial data for Blair's report. In this sense, Bitumount was leagues ahead of any of its predecessors and pivotal to subsequent policies and projects.

* * *

Sidney Blair's "Report on the Alberta Bituminous Sands" was finished in December 1950 and presented in an edition of fifteen hundred as a Sessional Paper of Alberta in May 1951.[46] Blair's conclusion was forthright. He stated that "the Bituminous sands can be mined and the bitumen processed by established methods," in his introductory sentence. Blair estimated the mining, separation, refining and transportation costs to be $3.10 per barrel of crude oil delivered to the Great Lakes ports. This compared to the $3.50 per barrel that comparable oil distillates were receiving. Blair argued that a profit of 40 cents a barrel was available as of December 1950 and he concluded that the amount was sufficient to justify plant and transportation investments. Moreover, since the necessary pipeline and refining plant could be linked to conventional reserves in the Edmonton area, the development cost of the oil sands would not be particularly large. If this series of conclusions was accepted, Blair argued, there were no barriers to oil sands development.[47]

There were, however, three caveats to Blair's study. First, he admitted that further studies must be undertaken of the technology of separation and of the magnitude of oil sands deposits. Second, he emphasized that the larger the scale of the plant, the more readily would the economics of production be realized. Blair claimed that a 20,000 barrel-a-day plant was a minimum size. Indeed, his calculations of mining, separating and refining costs were based on this minimal size. Third, he noted that policy regarding the place of oil sands in the larger production of oil in Alberta must be reviewed, although he was vague as to what revisions he sought.[48]

Blair's report surveyed both the Bitumount hot-water plant and the alternative techniques of cold-water and "fluidized solids" separation developed by federal government researchers. He described the operation at Bitumount, not surprisingly, as the practical manifestation of Karl Clark's research for the Alberta Research Council:

The separation of the Alberta bitumen from sand with the aid of hot water has been extensively studied. The principles by which bitumen can be efficiently separated from quartz sand with the use of hot water are thoroughly described in the publications of the Research Council of Alberta. . . . The research has shown that when an intimate mixture of bituminous sand and water is heated the oil separates from the sand and disperses into small flecks of variable size that lie among the sand grains.

The oil in this pulp can be collected by flooding it with an excess of hot water. The larger flecks become readily associated with air and water vapor and float to the surface of the water where they form a fairly stable froth that can be conveniently removed. It has been found that the clayey material in the bituminous sand is closely associated with the oil dispersion and its possible collection. The presence of a certain amount of fine materials has been shown to aid the separation.

. . . With suitable operating conditions a bitumen froth can be produced that does not carry more than 5% of mineral matter and it usually has around 30% of water. The operation can be so directed that the primary froth accounts for about 90% of the total bitumen and it has been shown that an additional 5% can be efficiently recovered by the further treatment of the material in suspension. Thus it is possible to attain a total yield of 95%. . . . The work at Bitumount has demonstrated that the hot water separation process can be carried out continuously and can be placed on a routine operating basis on a large scale.[49]

The rest of the report contained other appraisals of the possibilities of the other separation techniques. But its main thrust was to emphasize the practicality of the hot-water method, its adaptability into large-scale production and the economic effectiveness of the separation process.

In sum, Blair's report indicated the likelihood of commercial development. Privately he told J.L. Robinson of the Board of Trustees that development was possible rather than probable. But he warned that the Bitumount plant would be required for much further testing by government and possibly industry. Indeed, he argued that industry would certainly wish to have further corroboration of his own results. He argued that a public presentation of information to the entire oil industry, as well as further research on several fronts, would be necessary to secure commercial development. Blair offered himself as a consultant for Alberta in order to sustain the "momentum" he thought the oil sands had gained.[50] Blair's argument and his offer were accepted and two further developments central to the oil sands occurred.

One was the public symposium, the Athabasca Oil Sands Conference, held in Edmonton in September 1951. At this conference were some 140 delegates from oil companies, principally American, and other businesses, as well as from government agencies. The programme even included a visit to the Bitumount site. Papers on the technological aspects of the oil sands were presented, generally sustaining the Blair summation of research, as well as urging continued work on hot and cold water, fluidized solids and the more experimental *in situ* recovery systems.[51] But perhaps the most important presentation was that of N.E. Tanner, Alberta's Minister of Lands and Minerals.

Tanner encouraged oil sands development but warned that legitimate developers rather than profiteers would be granted leases. He purported to announce definite policy regarding the place of the oil sands as part of the total Alberta oil marketing system, but his real claim was that oil sands developers would have their product and its marketability assessed only when serious development was undertaken. This probably signalled what the political scientist and oil sands analyst Larry Pratt has suggested was the refusal of the government to admit oil sands production into the existing prorationing scheme by which all Alberta oil production was marketed. Tanner did not make this point clearly in offering only general encouragement to oil sands developers and in warning that existing producers had prior marketing rights. Tanner held open the possibility of revision to government policy once oil sands development appeared to be nearing production.[52]

No assurance of marketing and no solid economic basis for the construction of a necessarily large-scale oil sands separation plant could be provided in 1951. If Alberta had been worried about the oil sands as a costly investment in 1946 and 1947, the enormity of oil discoveries from 1947 onward reduced the economic motive behind pressure for oil sands development and reoriented the provincial government toward the marketing of massive and assuredly-productive oil fields. There was, after

143

1947 and 1948, no pressure to develop the oil sands. Canada had a bountiful and cheap supply of conventional oil and Alberta had an assured, royalty-producing natural resource. There was nothing that the oil sands zealots could say about strategic need to counter this fact. (See Tables 5.2 and 5.3.)[53]

Table 5.2
Alberta Petroleum Production 1941-1951 in million barrels)

Year	Alberta Production
1941	9.9
1942	10.1
1943	9.6
1944	8.7
1945	8.0
1946	7.1
1947	6.8
1948	11.0
1949	20.2
1950	27.6
1951	45.9

Table 5.3
Crude Oil Wellhead Prices Per Barrel

1948	—Average $3.22
1950	—Redwater $2.73 Leduc $3.05 Golden Spike $3.03

Sources for both tables: Petroleum and Natural Gas Conservation Board of Alberta, "Alberta Oil Industry," 1948, 1951

A second—and private—response to the Blair study was even more discouraging to the hopes of oil sands backers. The British firm, Anglo-Iranian Oils (later British Petroleum) had been approached by Alberta's Agent-General in London, R.A.

McMullen, to examine opportunities in Alberta after the war. The hope of the Agent-General and the Economic Affairs Minister, A.J. Hooke, was that major redirection of British capital to Canada could be made in light of Canada's desire for foreign development capital and Britain's need to repay its war-time loans. Like so many hopes for a revival of the old imperial link, little came of this.[54] But one result was that the attention of Anglo-Iranian Oils was directed to the oil sands.

Anglo-Iranian was interested enough in finding alternative supplies to its Middle Eastern sources to send a team of researchers, led by Dr. D.A. Howes, to examine the oil sands. Howes's "Report," dated 20 June 1951, was very much a counterweight, both in conclusions and scope, to Blair's study. Although he had effusive praise for Bitumount as a test plant, for Blair's work and co-operation as well as that of Alberta officials in London and Edmonton, Howes was not convinced that oil sands development was feasible, let alone likely. As he summarized his information, "oil production from the bituminous sands is quite uneconomic at this time."[55]

Howes argued that the deposits in most areas were too variable in quantity to support sustained mining development. Only the Mildred and Ruth Lakes area contained sufficient quantities of good quality sands to sustain intense production. Here the impressive reserves could yield up to 840 barrels. (Blair had contented himself with references to the numerous reserves of 100 million barrels, but did not provide a detailed estimate as to their number.) Since surface mining was necessary, Howes thought that costs would be much higher, claiming that even large scale production would result in a net loss on each barrel of oil. Where Blair suggested that crude oil could be produced and transported for $3.10 a barrel, Howes argued that it could not be produced for less than $3.80 a barrel. A comparison of their respective cost breakdowns indicates their divergent views:

cost area:	Howes Report ($/bbl.)	Blair Report ($/bbl.)
mining	.59	.55
processing (separation and refining) ...	2.19	1.525
transportation	1.02	1.02
Total	3.80	3.095

Whereas Blair thought that $48.6 million would build a plant, Howes estimated it would cost $88 million. Thus, financing as well as production facilities (due to less than 100 percent efficiency in the separation process, desulpherization costs, the varying quality of bitumen) would be far more expensive than Blair had allowed. Finally, Howes argued that the value of the product at the Great Lakes market was more like $3.06 a barrel than the $3.50 Blair claimed. This was the case because of the peculiar quality of the oil produced from bitumen. It would sell at much less than the rates for Leduc or Redwater crude, which sold at $3.27 and $3.10 a barrel respectively. While Blair had suggested that the bitumen could be refined into a high quality crude, Howes scotched the notion on technical and economic grounds.[56]

The crux of the problem was addressed by Howes in his summary of the economics of development. Alberta's production levels, now enhanced by recent discoveries, met about one-third of Canada's 350,000 barrels per day need. But far cheaper foreign oil made up the bulk of Canada's supply, and eastern Canada was not likely to forego this cheap supply (Venezuelan and Middle Eastern oil comprised this supply) for higher priced Alberta crude, let alone for very expensive oil sands oil! The production of tar sands oil could be won only at the expense of conventional Canadian or foreign oil, and neither of these sources should be foregone. Moreover, Alberta, despite Lands and Minerals Minister Tanner's equivocal private view (to which Howes had access and referred), was not willing to let oil sands oil into the prorationing scheme which would have guaranteed a market for the oil sands.[57]

In sum, policy at both the provincial and federal levels, as well as economic rationality (the two criteria were distinguished), made it clear that expensive tar sands oil would not be wanted at present. Howes's report, then, was exactly the critical appraisal that Sidney Blair had expected from the oil industry and the skeptical estimation that the government would have to confute. While Howes's conclusions must have disappointed both Blair and the provincial government, they had laid out the conditions which development proposals must meet. If Blair had shown the technical feasibility of oil sands separation, Howes had shown the continuing economic and policy drawbacks to development.

Moreover, Howes gave the province very useful material in preparing for the Oil Sands Conference; it was no wonder that no sense of impatience betrayed the province during the conference or after. The government had already faced the realistic perspective of D.A. Howes.

The province's involvement in the construction of an experimental plant at Bitumount reveals four general points about the evolution of oil sands as a possible source for crude oil.

First, the Bitumount plant was the most successful operation up to that time. It was a superior operation compared with any of the succession of Abasand plants or with Fitzsimmons's operations. This success, however, was not unmitigated. The Bitumount plant was much more costly than planned, it took far longer to construct, and it operated for only five weeks in 1949. Nonetheless, sponsored by a government that was cautious about the project and its costs (the $1 million compares with a 1949-50 Alberta budget of $53 million), and willing to entrust it only to experienced oil sands engineers, Bitumount had delivered up the technical information that everyone sought.

A second point about Bitumount is that the plant ended up being operated by the government despite the government's own distaste for such "socialistic" schemes. In this regard it was run exactly as Abasand, even down to the fact that the private partner in the venture was unable to continue in the project, and yet that partner retained considerable holdings in the oil sands which would eventually pay off. The willingness of the provincial government, like the federal government before it, to carry out the project is an indication of the assistance which government was willing to grant business in order to support commercial development.

This co-operation indicates the contradiction between the ideological bent of the province and the reality of development in the post-war period. Despite its ideology of free-enterprise, the province was forced to support both the experimental work at Bitumount and grant concessions to the potential private developers in order to sustain interest in the oil sands. Alberta's push for development took priority over its commitment to private leadership.

Such support also meant that government regulation was crucial to the development of the oil sands. The third lesson of the Bitumount experiment is that policy as well as encouragement and aid remained central to the response of private enterprise to the oil sands. The issue of prorationing was pivotal to the decisions of various business firms in the fifties and sixties. It is to be regretted that the public record is so sparse on the thinking behind the government's hesitation to grant prorationing privileges to oil sands development. But the brute fact is that policy, as much as technical or economic criteria, affected the pace of development.

Finally, the issue of development was dependent in the post-war period not upon technology but upon economics. It was the impact of the price of oil and the market for oil which stalled the development of the oil sands, for pricing and marketing lay behind the government's hesitation over policy in the fifties and early sixties. The intrusion of market factors, and not a cabal of producers or politicians, inhibited the projects for oil sands development. Whether such market factors should be important is an economic and political issue, but the fact that they were crucial cannot be forgotten if the evolution of the oil sands industry since 1945 is to be understood. In 1950 the Province of Alberta had a $46 million surplus (it had only just begun to enjoy such luxury in 1948) which gives a good indication of the absence of any serious pressure to develop the oil sands once conventional deposits began to produce so much oil and so much government revenue. Similarly, the expansion of Alberta output to 45.9 million barrels, 96 percent of Canadian output and nearly 57 percent of Canadian supplies, represented a remarkable increase of output (see Tables 4.1, 4.2 and 5.2). Finally, in the period of the fifties, world oil stocks far outstripped demand despite the best efforts of consumer and industrial users in North America and western Europe. For the immediate future, the oil sands again were unneeded.[58]

It is instructive to compare the tar sands project with a parallel proposal to develop "synthetic" oil from the Colorado oil shales. Begun out of the same strategic concerns which drew Ottawa's attention during World War II, but continuing into the post-war period, the American Bureau of Mines investigated the extraction of oil from the oil shales as well as coal hydrogenation and gas synthesis. The continuing American government involvement emerged from an immediate post-war concern that world oil supplies were neither secure nor voluminous, and from intense regional pressure to develop resources (controlled by the federal government) on lands in the western states. Appropriations totalling $50 million and the construction of pilot plants left the Bureau of Mines and western congressmen favouring the public investment of up to $8 billion on oil shales and coal and gas synthesis plants. Intense effort by the National Petroleum Council and politicians from oil-producing states challenged the commercial viability of production from the oil shales, which the Bureau of Mines thought was economically-feasible. The American case is much like the Canadian one, with the goal of regional resource development running against commercial tests of feasibility. Like Alberta's, the American government sustained its interest in experimental oil production through calculations of non-economic concerns, but ultimately bowed to commercial arguments in a period of at least short-term petroleum abundance.[59]

VI: The Riddle
of the Tar Sands:
Some Conclusions

Tar sands development after 1951 has continued to reflect the Sidney Blair-Douglas Howes debate. The lure of development remained powerful. It has resulted in the remarkable projects of Suncor and Syncrude. But the problems of economic production and technical success bedevilled business and government planners alike. Both were bound up even more tightly than before in the resolution of the problems created by development of the tar sands.

* * *

In the post-1945 period, when the Alberta government sought to diversify the provincial economy, it returned to developing its natural resources armed with full jurisdictional control over them. While the Bitumount project was one of many means by which the province hoped to carry out its aims, it seems clear that the discovery of Alberta's conventional oil reserves, which started in 1947 and 1948 with Leduc and Redwater, was significant in limiting the importance of the Bitumount undertaking. Both Bitumount and Sidney Blair's analysis presaged the commercial exploitation of the oil sands. Both showed that such exploitation was really a matter of converging economic and commercial circumstances, and that the technical problems could be solved. This convergence, however, was to take from 1951 to 1964 because the world, including Canada, basked in the exploitation of remarkably cheap, reasonably secure, and apparently inexhaustible oil wells.

The cheapness of conventional oil in the period of abundance after 1947 can be seen by a measure the Canadian Petroleum

151

Association had provided. From 1947 to 1971, the Consumer Price Index in Canada increased from a base of 100 to 202. In these years, the price of crude oil moved from $16.00 per cubic meter ($2.54 per barrel) to $17.22. From 1971 to 1981, however, while the Consumer Price Index increased 2.4 times, from 202 to 480, the Canadian price of a cubic meter of oil increased nearly seven-fold to $117.70 per cubic meter ($18.62 per barrel). As far as supply was concerned, established reserves in Canada alone increased seven times between 1951 and 1971 from 218.8 million cubic meters (1,376.3 million barrels) to 1,584.4 million cubic meters (9,965.9 million barrels). (In 1983, proven conventional reserves were stated by the Geological Survey of Canada to be 1,173 million cubic meters or 7,378.2 million barrels.) Canadian production expanded ten-fold from 1951 to 1971 to 75 million cubic meters (471.8 million barrels) per annum in 1980.[1]

While the price of oil remained remarkably constant for more than two decades after 1950, the importance of oil as an energy source grew. Petroleum supplied approximately 30 percent of Canada's primary energy needs in 1950; by 1970 oil contributed 48 percent to total primary energy consumption, which had itself increased by more than 150 percent.[2] Oil therefore became an increasingly important commodity in a world where energy consumption had become crucial. The increased reliance on oil and concerns about security of short term supplies and long term reserves meant that the tar sands remained an energy siren whose lure would increase if supplies were threatened or prices rose.

Despite the oil market of the fifties and sixties, potential investors and experimental studies of the oil sands continued and at a quickened pace even in the fifties. Perhaps the most telling gauge of this is the number of test-holes drilled in the oil sands region during the periods of exploration. In the entire period of early drilling, from the 1880s to the 1920s, 34 wells were sunk. During the period of careful federal-sponsored core-drilling by CM&S, 291 test holes were sunk from 1942 to 1947. After 1947, about 100 holes were drilled in the 1952 to 1954 period. But from 1954 to 1963, 1,300 wells or test-holes were drilled by about 30 organizations. Even as conventional oil discoveries in Alberta were made, investigative work on the oil sands actually increased.[3]

Such veterans as Karl Clark and Sidney Blair kept track of the activity which continued in the fifties. They noted the exploration of serious oil companies like Socony-Vacuum, Sun Oil and Calvan in the oil sands. While these companies did not state very much about their intentions or hopes, unlike a couple of more brash firms, their presence at least indicated that the oil sands remained a possible site of development.[4]

A number of schemes to develop the sands were also presented to the province during the mid-1950s, when the provincial government tried to sell Bitumount. The government listened politely to several offers but refused to force the pace of development. The coolness of the government was indicated by its reaction to the Oil Sands Limited proposal to purchase Bitumount plus leasehold lands for $500,000. The province refused such an offer from Oil Sands Limited as it was then constituted.[5] Similarly, the government was pleased enough with the interest in 1954 of a Calgary firm called Can-Amera Oil Sands, headed by S.M. Paulsen and backed by a number of wealthy Americans from the du Pont and Ford families. But, though the company was granted the use of Bitumount in 1954 when it used the plant for tests (apparently contracting out the operations to a major oil firm), its prospects soon gave cabinet ministers, like Gordon Taylor, considerable unease. Can-Amera was thereafter discouraged by the province.[6]

Amidst this minuet of interested groups, Oil Sands Limited had managed to reconstitute itself late in 1953. The Great Canadian Oil Sands consortium was formed in 1953 from Abasand Oils (comprising both the partners in Ball's Abasand Oils), Canadian Oils Ltd., Champion's Oil Sands Ltd. (with major financial backers from central Canada), plus the prevailing force of the dynamic American oil firm, Sun Oil Co. of Philadelphia.

Sun Oil Co. was an aggressive independent company, involved in the "integrated" operations of exploration, production, refining and marketing oil and natural gas. The company's key characteristics help to explain its interest in the tar sands. Sun had consistently developed innovative refining and marketing processes which had allowed it to hold its own in American and Canadian markets against larger competitors from 1900 onward. Sun had developed refining techniques to handle the heavy-

153

gravity crudes of east Texas in the 1920s. Like Universal Oil Products, Sun was aloof from the "patent club" orchestrated by Standard Oil to share all refining innovations in the inter-war period. Moreover, Sun was managed by more consistently market-oriented executives, its proprietor-owners the Pew family, than most of the integrated American producers. Sun resisted both industry and government-led programmes of pro-rationing oil production even when the majority of the integrated American companies agreed in the late twenties and thirties to participate in supply-management schemes. By the forties and fifties Sun had come to terms with business-government co-operation in co-ordinating oil production and development. But the company does not appear to have been beset by the style of risk avoidance which John Kenneth Galbraith found to be the dominant mode of corporate management by the fifties. Rather Darwinian in outlook, experienced in innovative refining and production of oil products from heavy oil, and eager to find a massive oil supply to compete against producers with access to the huge Middle East fields, Sun was perhaps uniquely suited to take on the tar sands.[7]

After years of financial jockeying and negotiations with provincial and federal governments, the Great Canadian Oil Sands group contracted in December 1962 with the Bechtel Co. of San Francisco first to study and then to construct a large-scale commercial plant in the Mildred-Ruth Lakes deposit, north of Fort McMurray. By then, GCOS had spent $1.4 million on experimental work. The GCOS plant was built between 1964 and 1968. It was authorized by Alberta's Oil and Gas Conservation Board (predecessor to the current regulatory agent, the Energy Resources Conservation Board) to produce at an initial capacity of 31,000 barrels per day. The output of the GCOS plant has varied and expanded until its current capacity is 65,000 barrels per day. In comparison, Imperial Oil's Strathcona refinery in Edmonton, one of Canada's largest, is rated at 144,000 barrels a day. The first GCOS plant cost $220 million to complete. After many technical and financial vicissitudes, the GCOS plant, renamed Suncor following business reorganization in the late seventies, became a financial and technical success. Business historian Arthur Johnson argues in his study of Sun Oil that until the economic environment of the 1980s, the Suncor venture was not strictly profitable. Indeed, it then seemed that Suncor was the reflection of Sun's chairman Howard Pew and his view of business "challenges," and Canadian government co-operation than a prudent investment. Nonetheless, in 1984 Suncor reported consolidated or cumulative earnings of $1,460.8 million since its inception.[8]

The Suncor plant was followed by a second major oil sands venture, that of Syncrude. Syncrude was a project sponsored by a consortium of oil companies (including Esso Resources, Gulf Canada, Canada Cities Service, Hudson's Bay Oil and Gas) plus the Alberta and Canadian governments (and briefly Ontario, although the province later withdrew; in 1982 Ontario bought into Suncor). This joint public-private enterprise emerged from the 1973 shifts in world oil prices and the apparent ending of an earlier, wholly-private Syncrude project. Only after complex negotiations did the two levels of government put together the final Syncrude project acceptable to the oil companies in 1974. The plant, situated north of Fort McMurray near the Suncor operation, was completed after four years of construction, again mostly by the Canadian Bechtel Co. The cost of this second oil sands mining, separation and refining venture was, however, $2,200 million—ten times the cost of the first GCOS plant. Syncrude was licensed by Alberta's Energy Resources Conservation Board to produce 125,000 barrels of oil per day. This capacity has increased by some 10 percent, although technical problems have resulted in several long shut-downs; this has been the case at Suncor, too. Nevertheless, Syncrude is one of the single largest petroleum production plants in Canada.[9] Since Syncrude does not publish audited financial statements—its corporate owners report their share of Syncrude's costs and earnings in their statements—the Company's profitability remains unknown.

A number of points about Suncor and Syncrude indicate the web of factors shaping their development. The two enterprises were built only after considerable negotiations among participating business interests as well as the regulating, financing and taxing governments, federal and provincial. The plants were built when negotiations ensured financing, marketing and taxing regimes which were peculiar for the oil sands plants. Oil sands production, "synthetic oil," has been sold at world prices since 1978, despite a ceiling on "old" oil prices from 1973 to 1985. Very high production levels were guaranteed for the operators, as

Blair and Howes suggested, for production from Suncor and Syncrude has been exempted from the pro-rationing of production to which conventional Alberta oil is subject. The plants were not even planned until the completion of pipelines which provided a national and continental crude oil transportation system. The costs of the plants were extremely high even then. While Blair had thought that a large 20,000 barrels per day plant could be built for under $50 million, the original Suncor operation of 30,000 barrels per day production had cost four times as much to build. This accelerated inflation also flowed through to the Syncrude project. One constant in exploiting the tar sands has been that very large scales of production and extremely large amounts of money were needed to make the oil sands effective as a petroleum source, just as engineers and businessmen decided in the thirties and forties.

The most recent proposal for commercial development, the now-postponed Alsands project at the Fort Hills area north of Bitumount, reveals the continuing difficulty in resolving problems of regulation, production and economics for the tar sands. The Alsands group, led by Shell Canada, with participation by Petro Canada, Dome, Amoco, Gulf and Chevron, had become extremely cautious even when the price of oil more than doubled on world markets in 1979-1980 and when Canadian oil price policy guaranteed world prices for tar sands production. Despite tax concessions and marketing guarantees, the Alsands participants balked at the costs of a project which was estimated to run to at least $14 billion for construction of a 140,000 barrel plant. A recent analysis of the Alsands project by the Economic Council of Canada concludes that the project would indeed have been an economic liability, given the market and price conditions of the mid-1980s. In the absence of any subsidies, the supply cost of synthetic crude from Alsands would be approximately $315 per m³ ($50 per barrel) when the 1984 price was $240 per m³ ($38.15). Even with the offer of significant subsidies—through royalty holidays, deferred taxation, guaranteed markets—a tar sands plant was not considered a wise investment in the climate of the eighties, when oil prices and markets are highly unpredictable.[10]

The difficulty in producing and marketing products from the oil sands serves as a continuing inhibition to development. The role of government in co-ordinating and regulating the economic environment has been as crucial to the approaches of private enterprise to the tar sands since the fifties as the role of government in sponsoring research was in the pioneering era. Private business has always sought active co-operation and assistance from governments prior to proceeding with experiments or developments. Governments have accepted this role in the tar sands and other ventures, in order to diversify the economy and to search for healthy revenues for their other activities. Government involvement is also, of course, a consequence of Crown ownership of natural resources as well as government regulation of trade and commerce. In sum, there is a continuing Canadian pattern of asserting via government a public interest in economic as well as social and political matters. In a frontier area like the Prairies, developing resources which are part of a world market has imposed many problems. Governments have nurtured development and defended the public interest in these conditions. Governments are also called to act against dominant international firms which are capable of and willing to act to set production and pricing schedules and which have their own agenda for development. The links between government and business, then, which are found in the tar sands, result from the particular difficulties in producing oil from the tar sands as well as from the broader inter-reliance of government and business in economic activities that has marked Canadian economic development.

Regardless of the public's disposition towards particular policies or towards co-ordination of economic affairs by public and private enterprise, Canadians live in a society in which such inter-reliance has become the norm. It emerges from the Canadian dependence on such commodities as oil, even as local or regional identities or autonomy are asserted. The inter-reliance of this nation's governments and business in planning development is a pattern universal in the industrial-commercial world. Conflicting demands (commercial and strategic) of producers and consumers, provincial and federal taxation and regulating regimes, simultaneous private development and public ownership of resources, each has been raised sharply in the twentieth century in Canada. These facts have imposed more complications on resource development than merely the resolution of technical problems. In the case of the oil sands, no simple answers—economic, technical or political—provided, or yet provide, a short-cut to development.

A long era in which federal and provincial government enterprise led to private involvement marked the history of the oil sands from the 1870s to 1930s. A shorter and sharper period of federal and provincial government activity then led to private development in the 1940s and 1950s. These waves of public and private enterprise together may be said to have eroded the barriers which prevented the development of the oil sands as a petroleum source and as a commercial undertaking. The activity of federal and provincial researchers and private developers also resulted in the knowledge and confidence which resulted in the construction of the large commercial oil sands operation of the Great Canadian Oil Sands (now Suncor) and the Syncrude corporations, the development of the 1960s and 1970s.

The first waves of experimental activity made clear that the three problems of mining the oil sands, separating bitumen from the sand, and finally refining the bitumen into petroleum products had each to be dealt with. Moreover, from the beginning, researchers and speculators realized that the resolution of these problems required technical solutions which were also economically feasible. In other words, the technical avenues pursued were those which promised simultaneously to resolve economic barriers to development.

The development of the oil sands up to the point at which fully commercial exploitation was considered feasible, then, is the result of both public and private research and labour ever pointed toward commercial development. The Bitumount plant was and is the testament to the feasibility of commercial exploitation of the oil sands. The plant and site also perfectly represent the overlapping domains of government and business and of public and private development goals in Alberta's history. The shifts between the two governments and the different private developers can best be seen in oil sands development up to 1951.

In the early period of government research, the Geological Survey of Canada scientists operated from the premise that the oil sands were but a surface oozing of the immense reservoir of bitumen or crude oil underlying the Athabasca region. The geological theory of Bell and McConnell led a generation to expect that exploitation could result from the well-established lines of drilling for oil. Federal government interest from the 1880s to 1910s was encouraged by the hope that that exploitation would not be difficult. But federal participation in surveys and studies was also based on the ready acceptance by successive Dominion governments that the Dominion must assume responsibility for research as a means to aid the development of Canada's prairie and northern frontiers. Holding title to the region and its natural resources (from 1870 to 1930, thus even after provinces were created in 1905), the Dominion government took seriously its not unrealistic but premature hope for extraordinary agricultural and industrial development along the Canadian frontier. In this way, Canada laboured under the belief that national expansion would occur in a fashion similar to that of the United States, at once a feared rival and the model of transcontinental and commercial expansion and prosperity.[11] The first Geological Survey drilling work in the oil sands ended in a technical and ecological failure in 1897. However, the Dominion also presided over the delayed but still remarkable success of prairie settlement after 1900. Thus Ottawa remained willing to continue to conduct experimental work on resources like the oil sands.

Sidney Ells's work in the years up to World War I served a number of purposes. It was a critical check on the federal government's hopes for the oil sands. It led to an entirely new approach to the development of the Athabasca sands. Finally, it stood and indeed continues to stand as an object-lesson in the complexities which the oil sands pose to researchers and developers. Ells must be credited with some useful and even heroic effort in his surveys of the oil sands. Ells's studies with the Mellon Institute scientists, in particular, led him and other researchers to understand that the oil sands as such comprised the resource to be exploited. That is, Ells's work led to the conclusions that the oil sands were similar to other bitumen-and-mineral deposits reservoirs in North America and that the separation of the bitumen from the mineral matter constituted the only way that petroleum would be recovered in the Athabasca region. In this regard, Ells helped to orientate correctly the geological and engineering approaches to the Athabasca. Ells stimulated his fellow engineers, particularly his colleagues at the Mines Branch, to think about adapting ore-dressing technology to the separation of bitumen from oil sands.

Ironically, Ells's work had no impact on his contemporaries among the profit-seeking private developers in the early 1900s. These men continued to accept the somewhat misleading conclusions of Bell and McConnell about the traits of the bitumen deposits. Thus, a "generation" of private entrepreneurs experimented in the oil sands particularly during that period. This work typified the problems of premature activity on the part of business, for the efforts and money of a number of men and firms were wasted.

While Ottawa's determination to continue oil sands work was deflected by Ells's findings and by private activity, the province of Alberta emerged as the major participant in the oil sands in the 1920s. Goaded by the University of Alberta's President, Henry Marshall Tory, driven by its own concern to diversify and expand the provincial economy, and determined to gain control of natural resources and their revenues within provincial boundaries, the government of Alberta set into place institutions and initiatives which deeply influenced the oil sands' history. The province began to sponsor industrial research through the Research Council of Alberta, founded in 1919. It also began the negotiations in the early twenties that would end in the transfer of natural resources to provincial jurisdiction in 1930. Thus, the government and University of Alberta set about to develop and gain the benefits from the natural resources—the natural gas deposits, the coal reserves, and the peculiar oil sands—which promised to diversify and enrich the farming-and-ranching-based economy.

The Research Council's work on the oil sands was conducted under Karl Clark, the former Mines Branch engineer hired along with several other Dominion government researchers by Henry Marshall Tory in order to make the Research Council an effective agent of development. Karl Clark quickly became intrigued by the problem of the oil sands, pushing lines of inquiry established by Ells and the Mellon engineers and hoping to make a break-through in establishing the commercial value of the oil sands.

The Research Council's work on the oil sands under Clark must be seen as essential to the development of knowledge about the resource's exploitation. Clark applied the most promising technique, the hot-water separation of bitumen from the oil sands, and developed it through a decade of work. He arrived at something of a breakthrough—not a "discovery" but an application of engineering techniques to a new resource. The assiduity of his efforts in working at the site of the deposits and in demonstrating both the feasibility of separating oil and the complexity of that extraction process were pivotal to further activity by government and business.

Clark and the Research Council also demonstrated that Alberta, despite not yet administering the natural resources within its boundaries, was embarking on a serious and difficult quest of trying to diversify from an agricultural economy by using the state—university and government-sponsored research in this case. The research was similar to that conducted by the Dominion government when it had tried to shape national economic expansion. The research was also a reflection of the faith in scientific knowledge as a necessary as well as effective way to spur development. Still, Alberta attempted to build up the province in the areas of coal, natural gas and oil sands as well as other mineral and forestry research. This effort by the Research Council under the close supervision of the University of Alberta and the farmers' government of the 1920s (the Premier directly administered the Research Council) comprise an early if generally unacknowledged stage in the province-building goals of a society trying to broaden its economic base.[12]

In the twenties, the role of the federal government shrank. This reflected its own willingness to co-operate with the prairie provinces and their quest for greater autonomy in the post-war climate of strong western discontent. The federal government was still interested in oil sands development through its support of mining and paving experiments. Both were subordinate tasks, however. The federal role in prairie economic development may be seen to have meshed with the three prairie provinces' convictions in the 1930 transfer of natural resources administration (and revenues) to the provinces.[13] Tellingly, Ottawa kept the reserve in the Athabasca oil sands region and thus held on to an interest in the cause of oil security.

Not the least of the Research Council of Alberta's achievements in the twenties was to demonstrate to private interests that the oil sands comprised a valuable source of petroleum. Karl Clark argued convincingly that the chief commercial use of the oil sands would almost certainly be for motor fuels and oils. His studies provided a useful avenue of research for some of the more advanced workers to follow, like the Americans McClave and Ball, as well as highly instructive lessons for deluded oil sands drillers like Robert Fitzsimmons. Ironically, the Research Council of Alberta's work of 1929 and 1930 was a demonstration of the limits and possibilities of oil sands exploitation just as experimental private enterprisers were at last able to take up the work themselves.

These serious private interests arrived just as the world economy became embroiled in a decade-long depression which severely limited financial and commercial opportunities, among other effects. Fitzsimmons's International Bitumen and Ball's Abasand Oils were left to show that economic exploitation depended upon careful and expensive mining and separating equipment using the hot-water process. By the end of the decade, the possibility existed of marketing the products from the oil sands in the Mackenzie River valley region. But the economic environment of world depression and the particular complexities of the oil sands had destroyed Fitzsimmons's firm and severely crippled Ball's so that neither took advantage of the slim opportunities. If that was not enough, the continuing technical bogeys involved in separating and refining bitumen from the Athabasca sands did the rest of the damage to International Bitumen and Abasand.

The history of Abasand Oils during the forties demonstrated the continuing intractability of work in the oil sands and the lack of business interest at a time when government, at least, thought development of the resource was important. Ottawa's willingness to take over Abasand after April 1943 reflected larger strategic considerations. But federal involvement also served well to address the concerns of private enterprise, which was more than willing to let Ottawa take over development costs. An unfortunate series of misfortunes and bad decisions certainly limited the Abasand plant as a federal project. Ottawa's involvement enlarged knowledge of separation techniques and the geology of the area—telling points which again showed how difficult the exploitation of the oil sands was. The federal government's plant promised even greater technical achievements with the cold-water separation process when Abasand burned down.

From the embers of this fire in 1945, Alberta's own province-building ambitions may be said to have flared up. The province, in most exquisite contradiction of its free-enterprise ideology, found itself a partner with a private firm, Oil Sands Ltd. (successor to International Bitumen) in constructing a new oil sands extraction operation at Bitumount. The eventual take-over by the Alberta government of the whole operation and the success of plant operations in 1949, however unnecessary as a technical demonstration, stand out as a critical point in the history of the oil sands. Bitumount is the monument to the technical success and the economic and social goal of development Alberta had sought since the 1920s.

* * *

The pattern of joint public and private development stands out as the means by which the oil sands were surveyed, examined and experimented upon. Public as well as private enterprise went into the experimental oil sands plants built in the 1920s, 1930s and 1940s. This partnership also marks the triumph of commercial plants in the 1960s and 1970s. Joint public and private development must be understood as essential to the entire course of oil sands development. This raises another notable aspect of the oil sands. As this book has reiterated, exploitation of the oil sands involved so many variables of engineering and economics and even politics, that the success of development was always difficult. The hopes of Robert Bell about pools of bitumen, the too-quick optimism of Karl Clark about his breakthrough of the early 1920s, the frustrations of the confused Fitzsimmons operations and the problems of Abasand as a private and a public project each reflected this difficulty. Even the success of Bitumount was hardly conclusive. The different evaluations of its data by Blair and Howes show this, as do the decades of research since 1951.

The long search for technically-effective means to separate bitumen from oil sand and then to refine petroleum products that were marketable teaches another, subtler point than sheer intractability in processing oil sands. That point is that there was no "discovery" or breakthrough in the processing of oil sands. The initiatives of Ells at the Mellon Institute in the early 1900s and the labours of Clark at the Research Council in the twenties involved the application of technology to a particular resource. Clark's work pushed the technique of physical separation, common to mineral processing industries, to a higher level of effectiveness; he did not invent anything. In view of the occasional assertions about "discovery" of the hot-water process by either Ells or Clark, it should be kept in mind that both engineers were working on adapting existing technology. Other studies of the question of scientific discovery, for instance Michael Bliss's on insulin, have shown that even breakthrough achievements are the result of a combination of incremental advances and, often, the perception that a breakthrough has occurred.[14]

Karl Clark's hope that he was on the verge of such a breakthrough motivated him in the early 1920s. His efforts did lead to improvements, although he was just one of a number, including the American James McClave and the Englishman Ernest Fyleman (who filed the first patents), working on hot-water separation. But, as we have seen, even Clark's hopefulness turned to discouragement when, in the process of field-testing at Dunvegan in 1925 and at Waterways in 1929-30, good quality separation did not readily occur. In this regard, Clark can be seen as one of those researchers who, like precursors to his contemporaries who isolated insulin, never quite succeeded in getting "pure" or continuous processes working. Clark, perhaps infected by the same optimism which spurred post-war Canadians like Banting, Best and his University of Alberta colleague Collip, seemed to expect the same kind of uncovering of a process which would lead to a marvellous breakthrough to benefit society. Unfortunately, in the oil sands such a breakthrough did not occur—due to problems of commercialization and process development rather than to the approach and technique itself. Again, complexity emerges as a crucial aspect of the project.

The problems which beset International Bitumen and Abasand as both private and public enterprise serve only to underline the complexity of the oil sands. Technical factors were not to be overcome except through careful research and considerable expenditures. As the marketing conditions became more favourable in the late thirties and early forties, a slow shift toward effective technical operations can be seen—even though particular projects did not yet succeed in establishing commercial operations. The success that Bitumount represented was the long-term result of work conducted for more than thirty years by engineers, chemists and geologists and also the result of vastly increased demand for petroleum. Sidney Blair's report of 1950 showed that the oil sands were becoming an economically-necessary resource, which is why he estimated that the cost of production was favourable to development. The world and, especially, the Alberta oil glut through the fifties and sixties delayed the final shift toward viable production from the oil sands.

* * *

The Athabasca oil sands have sometimes been conceived as a hydrocarbon El Dorado. They have come to be viewed more realistically as a vital and yet costly petroleum source. If development has been achieved, it has been reached only by applying the technical ingenuity which was gained through arduous effort. Development also met political-economic criteria which must push any commercial enterprise in the contemporary world. The breakthroughs of the commercial developers, though, rested on the work of federal and provincial government researchers as well as that of private developers. The slow pace of development from the early part of the century to the 1950s was irritating to a wildcat entrepreneur like Bob Fitzsimmons and puzzling to a laboratory researcher like Karl Clark. It was neither to businessmen-engineers like Max Ball and Sidney Blair.

Technical, political and economic problems were involved in assessing the oil sands to the point where commercial development became feasible. The early investigations of the oil sands, and the surviving Bitumount and International Bitumen plants, represent the culmination of one stage in the development of the oil sands. The era of investigation into the oil sands is a test-case of the growth of the petroleum industry and energy policy in western Canada.

Map 2. Athabasca Oil Sands Area and Selected Plant Locations

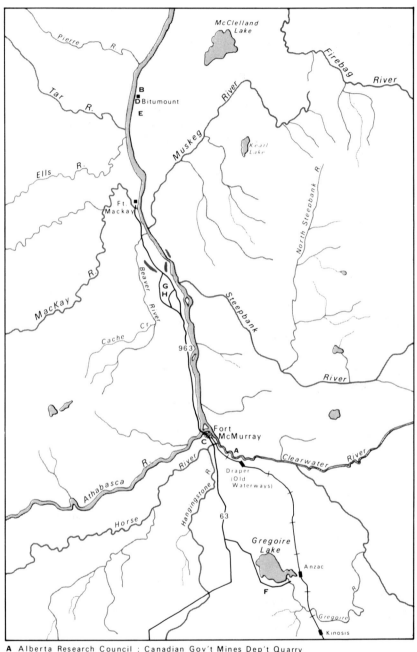

A Alberta Research Council ; Canadian Gov't Mines Dep't Quarry
B Bitumount : International Bitumen Co. (Robert Fitzsimmons)
C Abasand Oils (Max W. Ball); Canadian Mines Government Branch
D Bitumount : Alberta Government ; Oil Sands Ltd. (Lloyd Champion)
E Shell Canada "In-Situ" Test Project
F Amoco Canada "In-Situ" Test Project
G Suncor (formerly Great Canadian Oil Sands)
H Syncrude Canada

Railway
—63— Highways

Kms
0   8   16
0   5   10
        Miles

Appendix:
A Survey of Oil Sands Plant Operations, Present and Past

The preceding chapters have provided a narrative of the ways in which knowledge of and interest in the oil sands were translated into attempts at commercial development. If the discussions have explained the most prominent techniques, problems and advances which have accompanied research, they have not tried to present detailed accounts of the mechanics of plant operations or underlying principles. The detailed principles and practices of oil sands separation technology can be understood by two approaches to the subject. The first is to survey the major processes by which bitumen can be separated from oil sands based on current knowledge. The second is to examine the operations of experimental and semi-commercial hot water separation plants that led up to Sidney Blair's 1950 proclamation that oil production from the oil sands was feasible. (See Map 3.)

These approaches contain material that will be most interesting to technically-oriented readers. All readers might consider, however, that the subtleties of the technology - and the evolution of processes - help to explain why the application of scientific theory (in this case, that of hot water separation) was so intractable a problem for so long. Engineering problems required complex engineering experiments before any solutions could be found. This is to say nothing of the shifting economic factors which also conditioned the feasibility of extraction. The lesson from this technical review is that many problems had to be addressed before the oil sands could be exploited commercially.

167

The process of producing liquids from oil sand involves three steps (and current conventional terms will be used in the appendix):

1. Mining:
    to remove surface overburden and bring oil sand to the plant;

2. Extraction:
    to remove the very heavy hydrocarbon "bitumen" from the sand and water associated with it in its natural state;

3. Upgrading or Refining:
    to change the high density, high viscosity, sulphur-contaminated bitumen into usable final liquid products.

The mining and extraction steps can be replaced by a single step, *in-situ* extraction of the bitumen. This appears to be necessary when the overburden covering the oil sand is too great to permit surface mining. (These steps are summarized in Figure 1.)

The mining of oil sand has been an adaptation of standard coal strip-mining techniques. Upgrading is a technology borrowed from crude oil refining and has been known in its present form since the 1930s. Extraction and *in-situ* production are distinctive as to oil sands technologies. In these areas most of the technology has been developed to commercial standards in Canada.

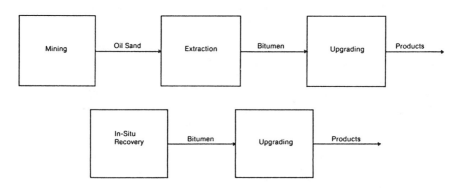

Figure 1.   Steps for Producing Commercial Products from Oil Sands

Extraction Processes

Many types of processes have been tried for the removal of bitumen from oil sand. Three methods have been proposed for surface extraction, feasible only when the overburden is less than 50 meters thick. These methods are:

1. extraction with water or steam;

2. destructive distillation;

3. solvent extraction.

If the overburden depth is too great to allow surface mining, an *in-situ* (recovery in place) method must be used. *In-situ* recovery combines the mining and extraction in an operation below ground. It involves pumping some form of energy underground to increase the temperature of the oil sands reservoir. Temperature increase makes the bitumen less viscous, causing it to flow away from the sand and eventually to the surface. Although a full scale plant has yet to be built using *in-situ* recovery, only 8 percent of the oil sand has less than 50 meters of overburden while 77 percent of the total sand is covered by at least 300 meters.[1] Many *in-situ* techniques have been proposed. Their success has been limited but testing continues.

Surface Extraction

1. EXTRACTION WITH WATER

a) Hot Water Extraction

Figure 2 is an enlarged picture of a typical oil sand deposit. The bitumen that is to be recovered from the sand makes up from 6 percent to 17 percent of the oil sand by weight. It acts as a glue that binds the sands together. Fortunately for the success of the water extraction methods, a thin layer of water surrounds each grain of sand and the addition of more water breaks the bitumen away from the sand at the bitumen-water interface.

"Athabasca bituminous sand is very amenable to treatment by the hot-water extraction method," noted Alberta Research

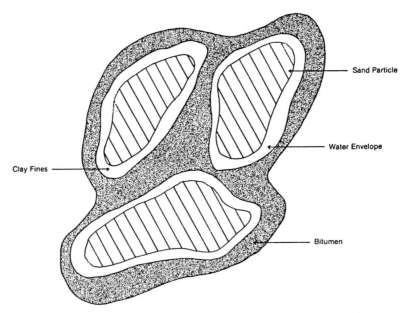

Figure 2.   Enlargement of Oil Sand Components

Council engineer, Karl Clark. "Indeed, the problem is not to find a workable extraction process; it is to work out the engineering details involved in handling the materials," Clark asserted.[2] The process of extracting bitumen with hot water was developed from the 1920s to 1940s. There are two basic steps:

1. Mix the raw oil sand with hot water, heat and agitate.

2. Separate the bitumen that has been separated from the sand and water by density difference. An oily froth containing the bitumen floats on the denser sand-water and is skimmed off.

As simple as these two steps sound, Clark's "engineering details" have presented problems that have been worked on almost continuously since 1924.

The major problem with hot water extraction is that the initial separation is not complete. The oily froth is contaminated with fine solid materials and water. The froth must be further processed. The rejected sand-water material must be further processed, for it contains 20 percent of the original inlet bitumen.

170

Figure 3 shows the modern hot-water extraction process that is used in the commercial Suncor or Syncrude operations. The processing steps are as follows:

1. Mix raw oil with hot water (80°C), low pressure steam, and caustic in a large rotating vessel called a tumbler. The steam adds energy to keep the slurry warm and aids in producing an oil froth. The caustic aids in rupturing the oil film that surrounds the sand grains and reduces the size of the tumbler that would be needed at the natural pH of the oil sand. The acidity of the initial separation vessel has been steadily increasing in processes used by oil sands pilot plants. This causes the pollution problems associated with the tailing water disposal to increase.

2. The slurry formed in the tumbler is screened to remove any large lumps of rock or clay not broken down. The slurry is then sent to a large tank called the primary separation vessel. This allows the solution to sit while the froth (bitumen, air, trace sand and water) floats to the top and is skimmed off. The bottom of the vessel contains a dense sand-water mixture which is pumped away as tailing. The middle stream (middlings) is removed for further treating to recover more bitumen.

3. The "middling" stream is treated by air flotation in secondary separation cells. This process involves floating the very small particles of oil left in the water by agitating and aerating the middlings. The bubbles of air introduced rise to the surface and bitumen attaches to them. The froth produced from this secondary separation is combined with the primary froth and is then further treated to remove sand and water. The secondary "scavenger" floatation is a common mining concentration technique and has been added since the early plants to increase the recovery of bitumen to over 90 percent.

4. The froth from the first and second flotation vessels is contaminated by fine solids and water. The froth will be settled out by gravity because of the viscous nature of the oil. To reduce the bitumen viscosity, a diluent light hydrocarbon is added. To increase the separation force this diluent-bitumen mix is centrifuged. This removes essentially all of

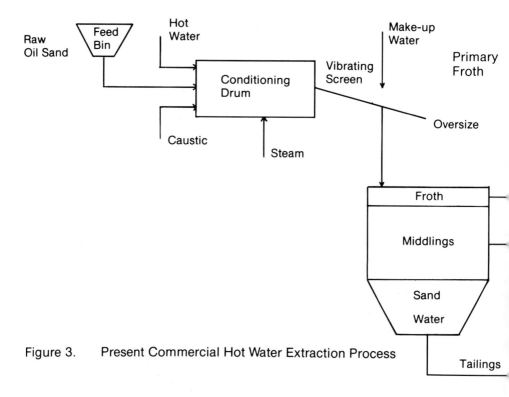

Figure 3.   Present Commercial Hot Water Extraction Process

the solids and water remaining in the bitumen. The diluent is then recovered by distillation and reused. The extracted bitumen is upgraded, or refined, into commercial products.

Several problems exist with the hot water separation extraction method that have yet to be solved:

1. Seven pounds of water are necessary to extract one pound of bitumen. This water is almost entirely removed as tailings and contains low level energy that is essentially lost.

2. The tailings water is caustic and contains much fine solids so it cannot be returned directly to the environment. This

172

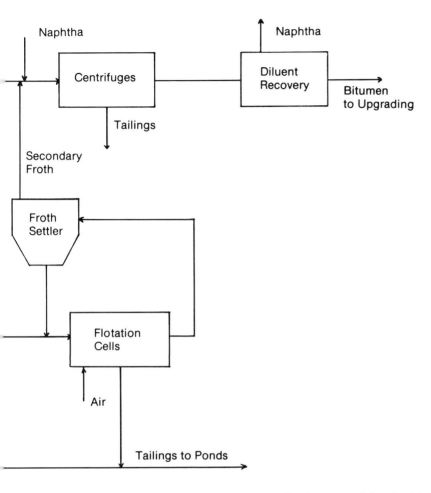

means that each plant has large "tailing ponds" to hold this water.

Several methods have been proposed as an alternative to hot water extraction in order to solve the problem of the energy loss. Extraction with cooler water has been proposed and tested.

b) Cold Water Extraction

A process similar to the hot water extraction was developed by the Department of Mines and Technical Surveys in Ottawa and tested at Abasand in the early 1940s.[3] The major differences between the hot water process and the "cold water" method are in the amount and temperature of the separator water, the initial addition of the diluent and the design of the first mixing vessel. Figure 4 shows the Mines Branch pilot plant extraction process.

The steps in the cold water process can be summarized as follows:

1. Warm water (25°C), diluent, caustic and oil sand are combined in a low discharge pebble mill. This replaces the tumbler found in the hot water extraction process but it is a similar slowly rotating device that contains pebble-size spherical steel balls. The energy provided by the hot water and steam in the hot water extraction process is in part replaced by the crushing and grinding action of the steel balls. This action helps break up the lumps of oil sand and in turn the bitumen-water-sand bond. The front-end addition of diluent (again a naphtha or kerosene-type liquid hydrocarbon) dilutes the bitumen so that it is less dense and can be separated by the next series of vessels. The pebble mill is a common device found in mineral ore extraction for reducing ore size and can be used in conjunction with a solvent for ore concentrating.

2. The discharge of the pebble mill is sent to a conventional ore rake classifier. This rake classifier is the replacement of the primary separation vessel and is a mechanical ore treating device. It allows dense material to settle to the deeper section of a sloped bottom vessel and a moving rake device to remove this solid material. More water is added at this stage. This will cause a bitumen overflow which is removed from the top of the rake classifier. This water must be metered very closely because too much water will cause excess sand to overflow the top of the classifier and too little water will cause the bitumen to be removed with the bottom sand product. The rake classifier has the advantage of removing the bulk of the sand as a solid and thus reducing the total amount of tailings water.

3. The diluted water-bitumen-diluent overflow from the rake classifier is then concentrated with a tray thickener to remove the water-fine sands mix from the hydrocarbon mix. A tray thickener is a tank containing several trays which slope

174

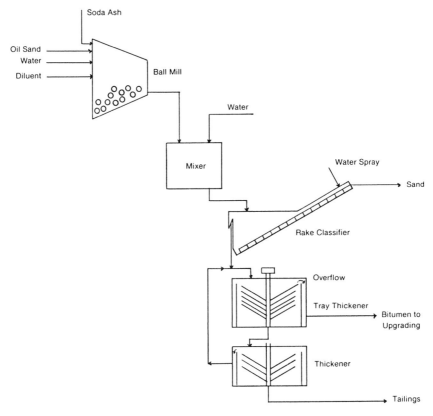

Figure 4.   Cold Water Extraction Process

toward the centre. The denser sand-water solution is removed from the centre of the tray with the aid of a scraping mechanism that channels the denser material to the tray centre. The less dense bitumen-diluent overflows from the edge of the tray and is ready for further processing. Again, tray thickeners are a common ore extracting device used in mineral recovery.

4.   The bitumen-diluent thickener overflow is then distilled to recover the diluent in much the same manner as the hot-water extraction diluent recovery.

Recoveries of 90 to 95 percent were reported[4] for the cold-water extraction process, which compares favourably with the hot-water method. In sum, there are three advantages to this process. It enables

1. reduced hot water and steam energy costs

2. removal of the bulk of the sand in solid form by conveyor

3. high recovery rates without extensive secondary recovery (flotation).

Drawbacks to the pilot process exist. These would be noticeable in the shift toward a commercial scale of operation. The drawbacks are:

1. Over twice as much total water circulation need than for hot-water extraction. 700 kg water/100 kg bitumen for the hot-water, 1500 kg water/100 kg bitumen for the cold-water.

2. More critical pH control (caustic addition) for efficient extraction.

3. Slight variation in oil sand quality, particularly the fine clay in the original water, greatly affected the recovery.

4. Excessive loss of expensive diluent with the tailings.

The research conducted into the cold-water extraction method is again going on at the National Research Council by B. Sparks and F.W. Meadus.[5] It is proposed as a secondary recovery technique to replace the flotation cells presently used in commercial operations. Also, Prof. E.Q. Neumann at the University of Toronto[6] has investigated new primary extraction techniques with colder water temperatures.

The problems of water disposal and energy losses associated with both the hot and cold water extraction methods have led to parallel research into alternate separation processes. This research started over fifty years ago and continues with increased vigour today. The major non-water extraction processes will be described next.

2. EXTRACTION OF BITUMEN BY DESTRUCTIVE DISTILLATION

If oil sand is heated to extreme temperatures (in excess of 500°C), the bitumen, like any petroleum, "cracks" or breaks into smaller hydrocarbon molecules. These smaller molecules vapourize and thus separate from the sand. The vapours can then be condensed to produce an oil residue. The residue is both less viscous and less dense than the original bitumen and can be considered partially upgraded. The general term for processes where this cracking takes place is destructive or dry distillation.

One of the first scientifically-rigorous pilot plants to test this method was developed in the 1940s by W.S. Peterson and P.E. Gishler[7] at the National Research Council in Ottawa. A diagram of their process is shown in Figure 5. The basic steps to this process are as follows:

1. Oil sand is reduced to lumps under 12 mm in diameter and forced into a fluidized bed reactor. (A fluidized bed is one where the catalyst or heat transfer material — in this case the solid sand phase — is made to bubble like a liquid by passing vapours through the solid material.) In this process the gases causing the fluidizing in the reactor are the hot vapours produced when the bitumen is cracked.

2. The gases leaving the reactor are then cycloned to remove any entrained fine solids. The liquid oil products are recovered by cooling or electrical precipitating. The product contains molecules from pentanes and heavier hydro-carbons. But, the final product has the physical properties of a very sour (5 percent sulphur), heavy (specific gravity of .95) crude oil. Sour gases of butane and lighter grade are also produced in the cracking process. These can be used to produce some of the energy required to raise the oil sand temperature high enough to cause cracking.

3. The sand left behind when the cracked bitumen vapourizes is coated with carbon molecules, coke. Air is injected into the process at the bottom of the reactor to carry the sand over to a burner. The air burns the carbon from the sand, heating it to well above the temperature required to crack the bitumen.

4. The gases formed from the coke and cracked gas combustion in the burner vessel are cycloned to remove fine sand solids. Heat from the gases is recovered to produce steam.

5. Some of the hot, clean sand is returned to meet the oil sand in the reactor vessel, providing the energy necessary to crack

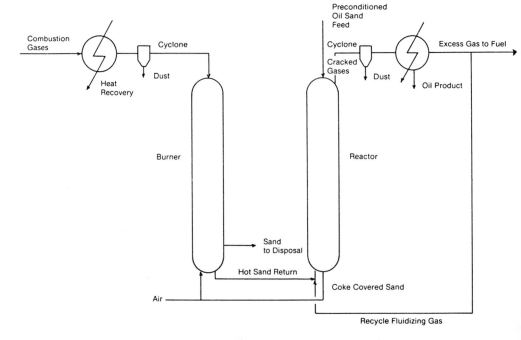

Figure 5.　　Extraction of Bitumen by Destructive Distillation

the bitumen. The excess sand is removed from the burner section as a solid with little carbon residue.

The destructive process is very similar to the upgrading process called "fluid cooking" that is presently used at Syncrude and in many refineries. However, instead of feeding bitumen or crude oil residue into the fluid bed, the process receives the oil sand directly.

The recoveries from the initial pilot plant study[8] were in excess of 80 percent and all the heat required for the destructive distillation was produced from the bitumen itself. The product contained a very high sulphur content and might require further upgrading to provide a better quality synthetic crude oil.

If this process could be made commercial, it would have at least two advantages over water extraction plants:

1.　No great amount of water is needed and none contacts the

bitumen itself. Hence fewer water handling and pollution problems would result.

2. The extracted oil is already considerably upgraded. If a market could be found, further upgrading may not be needed. If the high sulphur content was a problem, a simple upgrading scheme such as hydrocracking could be used to produce quality synthetic crude oil.

However, five problems would arise in a move to a commercial operation:

1. The combustion gases from the burner are high in pollutants and would require considerable clean-up before being discharged to the atmosphere.

2. Feeding large quantities of oil sand into the fluidized beds is a technology not well developed. Feeding solids into this type of reactor in similar parallel processes such as coal gasification has demanded the use of many small reaction vessels.

3. Recovery rates in the initial test case were lower than those for water extraction.

4. The sand in the oil sand varies greatly in size and could cause problems in keeping the bed adequately fluidized.

5. Considerable heat is lost with the rejected hot sand from the burner. Process energy efficiency could be greatly increased if the heat from this sand could be recovered.

Recently, Lurgi Canada Ltd.[9] tested a process almost identical to that of the Gishler process and is carrying out engineering studies for a 4000 tonne/day pilot plant. The firm expects gross recovery of bitumen to be between 92 percent and 95 percent with 80 percent of the bitumen being converted to the final synthetic crude product.

Also, research into destructive distillation in a non-fluidized bed is being conducted by W. Taciuk,[10] supported by the Alberta Oil Sands Technology and Research Authority (AOSTRA). This Direct Thermal Processor uses a horizontal rotating processor rather than the initial fluidized bed but the auxiliary equipment and oil product are very similar to the fluidized bed method.

3. OTHER EXTRACTION METHODS

Many types of extraction methods other than water separation and destructive distillation have been proposed. The two major alternatives that have been extensively studied are solvent extraction and oleophilic sieving.

a) Solvent Extraction

If oil sand is combined and agitated in a beaker with an organic solvent such as benzene, the bitumen is dissolved away from the sand almost completely. Recoveries approaching 100 percent are possible. Attempts to adapt this laboratory test into a commercial process have met with two major difficulties: the solvent is "lost" as a vapour in the initial mixing device or the solvent is "lost" in the rejected sand. One of the earliest and most extensive testing programmes into the solvent extraction process was conducted by Cottrelle in the late 1950s.[11]

The process postulated by Cottrelle and tested by Canada Cities Services Ltd., shown in Figure 6, consists of the following steps:

1. Mix the oil sand with a diluted solvent-bitumen recycle stream in order to form a slurry of oil sand and solvent.

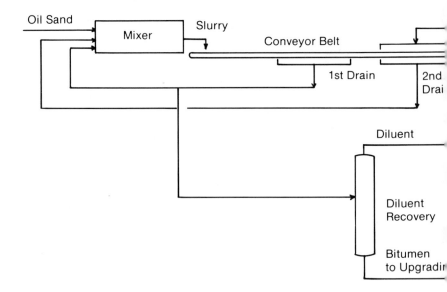

2. Drain the liquid away from the oil sand in a series of chambers as the oil sand slurry moves along a perforated conveyor belt. Wash the oil sand with pure solvent before each drainage stage.

3. Remove a portion of the drained solvent-bitumen extract and distill to recover the solvent for reuse and to send solvent-free bitumen to extraction.

4. Recover the solvent adhering to the bitumen-free sand by heating the sand with steam (steam-stripping) to vapourize the solvent and then dispose of the clean sand.

The main advantages of the process are that it uses no water and that it expels the sand as a solid.

On the other hand, several problems are predicted for this process when scaled up to commercial size:

1. Solvent losses in the rejected sand can still occur.

2. Solvent vapour losses will occur due to evaporation if the drainage stages were not adequately sealed.

3. Finding the correct type of solvent, flow rate of solvent, and

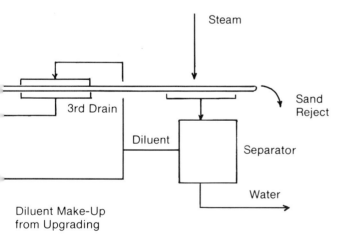

Figure 6.   Anhydrous Solvent Extraction of Bitumen

181

oil sand slurry bed depth for maximum recovery and optimum efficiency remains unclear.

The work on this process continues through preliminary economic studies now sponsored by AOSTRA.

b) Oleophilic Sieve

One of the newer methods being tested for extraction of oil sand uses an oleophilic (oil attracting) sieve that attracts bitumen but allows sand and water to pass through.[12] A slurry of bitumen-hot water is made in much the same way as in the hot water process. This slurry is passed through the oleophilic sieve. Bitumen is recovered by washing the sieve with solvent.

The main advantage of the process is the fact that only one-third of the water required for the hot water extraction method is required and this water is not caustic. This makes the treatment of the sand-tailings much simpler. The major drawback seems to be in building equipment to commercial size.

In-Situ Extraction

About 90 percent of the oil sand in the Athabasca deposit is covered by 150 meters of overburden and cannot be surface mined (Figure 7). If these reserves are to be developed, an in-situ (literally, a place below ground) method of extracting the bitumen from the oil sand must be used. The methods proposed and tested so far have been adaptations of petroleum production techniques used for the enhanced recovery of conventional crude oil. The common characteristic of all the methods is the placement and distribution of some form of energy into the oil sands formation. This energy is used to heat the formation, causing a reduction in bitumen viscosity with the temperature rise. In fact, a large viscosity reduction of bitumen with temperature increases was demonstrated in early oil sands drilling. The friction between the drill bit and the oil sand caused a temperature increase and liquified the bitumen. When the bit was removed, it was dripping with liquid bitumen. This convinced early oil sands explorers, like von Hammerstein and Fitzsimmons, that there were pools of liquid bitumen trapped in the sands.

The major methods of *in-situ* extraction have been developed since the mid-1950s following the initial development of similar techniques in conventional heavy oil fields in California and Texas. These techniques have not yet been used in a full scale commercial operation. However, they will have to be used for both heavy oil and bitumen recovery in Canada in the near future if Canada is to fully exploit the oil sands.

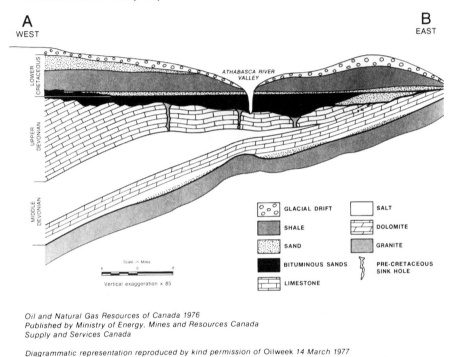

Oil and Natural Gas Resources of Canada 1976
Published by Ministry of Energy, Mines and Resources Canada
Supply and Services Canada

Diagrammatic representation reproduced by kind permission of Oilweek 14 March 1977

Figure 7.   Cross Section of Athabasca Oil Sands

1. Steam Injection

High pressure steam injection has been the most extensively tested method of *in-situ* extraction. Steam has several advantages. It carries a large amount of energy per unit mass. It is easily transported to the formation at high temperatures. The technology for generating steam is centuries old. The first large scale tests in the Athabasca area were conducted by Shell Canada Ltd.[13] in the mid-1950s. Initially only an alkaline water

183

solution and wetting agents* were injected into a well and then recovered as a water-bitumen emulsion from another well less than ten meters away. Shell soon discovered that high recoveries of bitumen could not be sustained with this technique unless the bitumen's viscosity was substantially reduced by temperature increase.

In the late 1950s, Shell again injected an alkaline solution into the oil sand formation and then followed this with steam. This procedure greatly increased recovery rates and throughout the early 1960s extensive tests were conducted which varied the injection rates, well distribution patterns, and stream pressure to get the best recovery efficiency. The method finally established consisted of the following steps:

1. Wells were drilled in a close space pattern (Figure 8) and were then fractured† to open channels through which fluids could flow to or from a well bore more easily.

2. Steam was injected into all the wells until the formation temperature was raised to an average of 175°C (about 100 days).

3. Steam continued to be injected into the exterior wells while extracted bitumen was produced from the centre well (about 800 days).

4. After a period of time, the steam injection was discontinued but production continued until the built up formation pressure declined (about 400 days).

Using this method of recovery, an average producing well would yield 200 barrels per day of bitumen and require 100 tons/day of steam. Recoveries of bitumen were between 25 and 30 percent. Three major drawbacks of this method exist:

1. Energy is lost within the formation. A very large percentage of the energy sent into the wells was lost by leaking outside the well pattern or to the surface. High pressure steam caused better heating and better bitumen recovery but also caused greater losses.

*   Chemicals similar to soap that help emulsify bitumen with water.

†   A common oil production technique which forces high pressure fluids into a formation to crack open channels followed by propellants to keep the channels open.

184

2. Costly fuel is used to produce steam. The steam generation used conventional, non-polluting fuel (natural gas) and the large fuel requirements make for poor economics.

3. Production expenses are, therefore, high compared to conventional crude oil. A plant producing 100,000 bbls/day of synthetic crude oil would require 3,500 wells (drilled every 3½ years) and 140 million pounds per day of steam.

Many test programmes have been conducted since the initial Shell test. Most recently, BP Canada and Petro Canada have completed construction and begun production based partly on

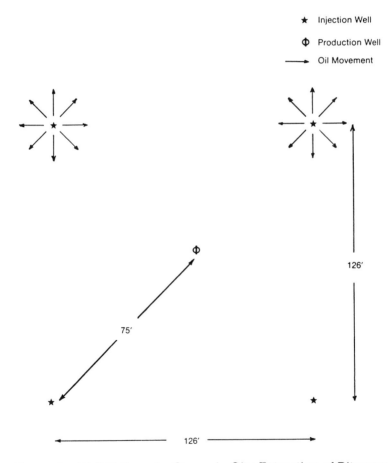

Figure 8. Well Pattern for Steam *In-Situ* Extraction of Bitumen

185

steam injection of an initial 192 wells at a site at Wolf Lake in the Cold Lake Oil Sands region.

An alternate method, however, solves some of the steam injection problems by burning the bitumen in the formation. This is described next.

2. *In-Situ* Combustion ("Fireflooding")

Extraction techniques that burn part of the bitumen underground to increase the temperature of the formation have also been studied since the mid-1950s.[14] This process has come to be known, graphically, as "fireflooding." The major advantage of this method over steam injection is that energy is produced in the formation itself instead of being generated by expensive, purchased fuel on the surface. Also, water processing requirements are greatly reduced. Furthermore, the bitumen combustion results in underground cooking of the bitumen. The oil

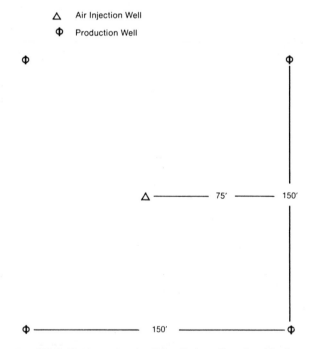

Figure 9.   Well Pattern for *In-Situ* Extraction by Underground Combustion

produced is up-graded to a lower viscosity and density than the original bitumen. Finally, the gases produced from the combustion products and vapourization of oil sand water pressurize the formation and help push the bitumen to the surface.

One process, proposed by AMOCO Canada Petroleum Ltd., consists of the following elements:

1. Multiple wells are drilled into the formation (Figure 10), consisting of an air-injection well surrounded by producing wells.

2. The central well is fractured to allow better air distribution into the formation.

3. Air is injected into the central well and ignited. Small amounts of water are added to help control formation temperature (200 days).

4. When a certain temperature is reached at the producing wells, air injection is stopped and the formation is allowed to depressurize (120 days). Significant amounts of bitumen are pushed to the surface by the gases (produced steam and combustion gases) during this period.

5. Water and air are injected into the formation again at a precise ratio. This causes a lower combustion temperature and turns the water to a steam that spreads ahead of the burning front pushing the hot bitumen ahead of it.* The air and water cannot be injected too quickly or the coke left over after burning the bitumen will form too quickly and plug off the formation. This last phase lasts from 150 to 300 days.

Twenty years of testing have improved well spacing and patterns, initial air injection rates, air/water injection ratios and injection pressures. The average producing well would make 200 barrels per day. Recoveries of 35 percent are realized if the gases produced from the coked bitumen are included. Seven percent of the bitumen in the formation is burnt.

---

* This procedure is called Combination of Forward Combustion and Waterflood (COFCAW) and has been used to increase recoveries of conventional crude oil.

In spite of the encouraging results, major problems still exist with *in-situ* combustion:

1. The combustion front does not distribute to the outside producing wells uniformly and may, in fact, head in an undesirable vertical direction. This may result in the warm bitumen not reaching the depth to which the well has been drilled.

2. The formation can be plugged by the coke produced in the combustion of the bitumen.

3. The pushing action of the water-air combustion may push oil sand ahead so quickly that the well plugs off.

In spite of these difficulties, this type of *in-situ* recovery is approaching economic feasibility. Research is being conducted by AOSTRA and many major oil companies. The BP - Petro Canada Wolf Lake project combines steam injection to isolate bitumen with combustion ("fireflooding") to ensure the flow of bitumen to the surface. Initial production of up to 7,000 barrels (an average of 35 bbl. per well) per day is reported and the process is commercial given special taxing and royalty relief.[15]

3. Other In-Situ Processes
Two alternate underground energy generating proposals that have not been as extensively tested are underground nuclear explosions and electrical induction heating.[16]

The use of nuclear energy would involve exploding a small atomic bomb below a deep oil sands formation (Figure 10). It is speculated that the explosion would form a large underground cavity and that the heat released would upgrade the oil sand to a less viscous oil product. This upgraded oil would then drop into the cavity and be pumped to the surface by conventional oil recovery techniques.

Some theoretical study of this project was done in the late 1950s and early 1960s, although pilot plants have never been conducted. They seem unlikely given Canadian public policy towards underground nuclear explosions. Also, considerable doubt exists that enough energy would be liberated by the explosion to increase the oil sand temperature sufficiently for economic recoveries.

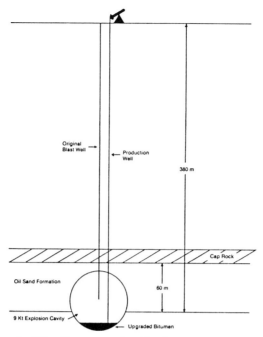

Figure 10. *In-Situ* Recovery by Nuclear Detonation

Electrovolatization is another method of raising the temperature of the bitumen by placing a large voltage potential across the oil sand itself. This causes the bitumen to become an electrical conductor. Because bitumen is usually such a poor conductor, the current forced through the bitumen will heat and actually vapourize it. The oil product, when condensed, is considerably upgraded, more so than any other *in-situ* method.

The limiting factor in this method seems to be the expense of generating the electricity at the surface. The expense of electrovolatization alone would be $100/barrel of upgraded crude oil but the possibility exists that this method could be combined with either steam drive or combustion to improve the efficiency of these methods. Research is presently being conducted into this method at the University of Alberta.

Presently, five active pilot projects are being conducted in the Fort McMurray area and a commercial plant could be in place before the year 2000.

189

Map 3. Select Athabasca Oil Sands Enterprises, 1900-1950

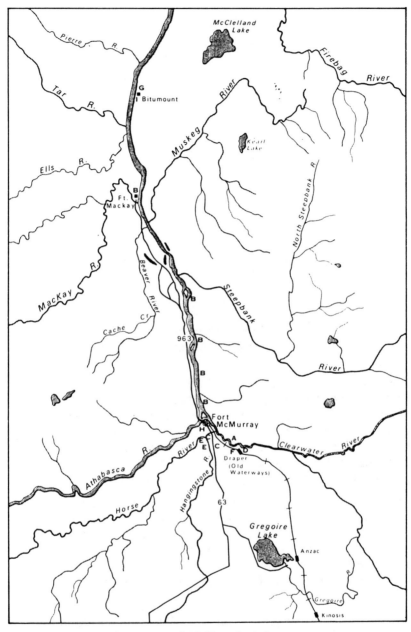

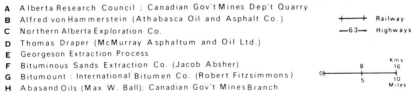

A   Alberta Research Council ; Canadian Gov't Mines Dep't Quarry
B   Alfred von Hammerstein (Athabasca Oil and Asphalt Co.)
C   Northern Alberta Exploration Co.
D   Thomas Draper (McMurray Asphaltum and Oil Ltd.)
E   Georgeson Extraction Process
F   Bituminous Sands Extraction Co. (Jacob Absher)
G   Bitumount : International Bitumen Co. (Robert Fitzsimmons)
H   Abasand Oils (Max W. Ball); Canadian Gov't Mines Branch
I   Bitumount : Alberta Government ; Oil Sands Ltd. (Lloyd Champion)

⊢ Railway
—63— Highways

        Kms
0   8   16
⊢───────┼───┤
  5   10
       Miles

Historic Plant Developments

The early history of oil sands development involved both laboratory and field research from pilot plants. This section will assess the major plants of the 1920s, 1930s and 1940s that sought to develop a commercial process in the Fort McMurray area by applying the principle of hot or cold water separation. The plants reviewed are those of the Research Council of Alberta in Edmonton and Waterways, International Bitumen Company at Bitumount, Abasand Oils Ltd. on the Horse River, and the Government of Alberta at Bitumount. Each will be examined for their major pieces of equipment, for their type and scale of operation, the success of the extraction method, and the technology learned from the plant.

A. Research Council of Alberta - Dunvegan Yards and Waterways Plants

The separation of bitumen from oil sand with the use of water was first attempted by G.C. Hoffman of the Geological Survey of Canada in 1883.[17] He reported that the bitumen separated readily from the sand. The first extensive research into oil sand reserves, extraction and uses was conducted by Sidney C. Ells. In 1915, he carried out experiments on oil sand extraction at the Mellon Institute of Pittsburg, Pennsylvania, and concluded from these tests that flotation cells were best for bitumen recovery. His work was flawed, but aspects of his suggestions were incorporated in later work.[18]

Soon after Ells's first investigations, K.A. Clark, assisted by S.M. Blair, and D.S. Pasternack[19] started investigating the extraction and uses of bitumen for the Research Council of Alberta. The first pilot plant constructed to confirm laboratory conclusions undertaken from 1921 to 1924 was built in north Edmonton at the Dunvegan Rail Yards in 1924. A schematic diagram of this plant as rebuilt in 1925 is shown in Figure 11. The steps in this initial extraction procedure were as follows:

1.   The oil sand was fed through a set of rollers (size reduction) into a mixing and heating vessel combining it with hot water,

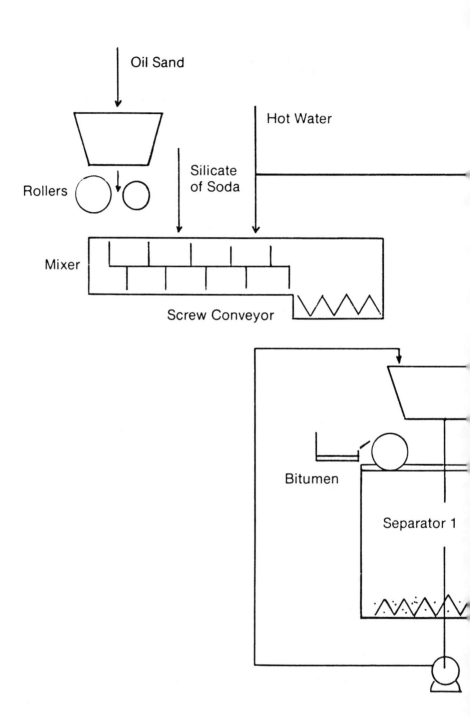

Oil Sand

Hot Water

Rollers

Silicate
of Soda

Mixer

Screw Conveyor

Bitumen

Separator 1

Figure 11.   Alberta Research Council Plant—Dunvegan Yards

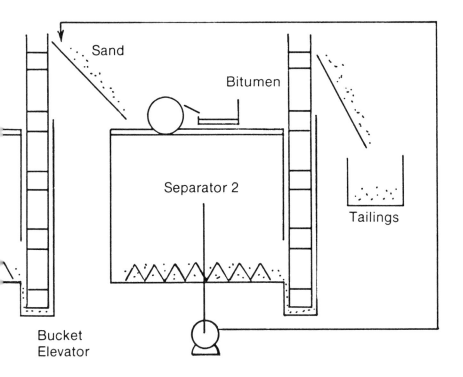

steam and a 3 percent silicate of soda solution. The silicate of soda was used as a chemical reagent to aid in breaking the bitumen-water-sand bond (like soap breaks a grease-solid bond).

2. The mixture was agitated by the blades of a steam jacketed clay and brick mill to produce a hot (85°C) thick mortar-like mixture.

3. The water-sand-bitumen was discharged through a screw conveyor into a funnel where considerably more hot water was added to dilute the mixture and wash the hot mixture into a separation tank.

4. The separation tank allowed the bitumen, water and sand to separate by density difference. The heavy sand was removed to a boot by another screw conveyor and carried away by buckets that were dipped into the boot. The buckets were raised through the hot water section of the separation box that was separated by a partition from the main section. The sand was washed from the bucket with a water spray and sent to a second separation box.

    The bitumen formed a hot froth at the top of the vessel and was removed by a slowly rotating steel drum that was dipped into the bitumen and then scraped into a trough for storage.

    The middling stream was pumped from the centre of the vessel to the funnel above the tank to help wash the oil sand from the mixer into the separation vessel.

5. The sand from the first separation cell just washed off the buckets was sent into a second separation vessel identical to the first. The sand discharged from the boot as a solid.

Several conclusions were drawn from the operation of this plant. First, the method of removing the sand with buckets was not satisfactory. The sand was not washed free of bitumen by raising it through the water. It actually picked up small droplets of the bitumen in the water and left with more bitumen that it contained in the initial separation vessel. Second, the almost-closed water. system (removing the sand with very little water) caused the fine clay particles present in the original oil sand water to build up in the recirculated water. This raised the water density to the extent that the gravity separation (bitumen raising through the water, sand sinking) was severely hindered.

Third, the second separation cell was ineffective and removed very little bitumen.

Finally, the bitumen produced from the separator froth was contaminated with large quantities of water (25 percent) and solids (3 percent) and would need further treatment to be used as a petroleum raw material.

These problems were worked on in the laboratory over the following four years. The design that resulted was built from Edmonton plant parts and new equipment on the north bank of the Clearwater River about a kilometer from Waterways in 1929. The new plant set-up is shown in Figure 12.

Although similar to the Edmonton plant, several important improvements were made to the second plant. The operation consisted of the following steps:

1. The oil sand feed was brought from a stock pile by a conveyor and bucket elevator. The conveyor passed under a barrier which sheared the large oil sand lumps and eliminated the need for the rollers found in the first plant. The buckets discharged into an automatic dumper which fed the sand into the mixing vessel in proportioned amounts.

2. The mixing vessel accepting the oil sand, water and steam was essentially the same as the Edmonton plant. The only change was a modification to the mixing blades to aid in the movement of the slurry through the tank.

3. The discharge of the mixing vessel was again washed with hot water through a funnel but a rotary screen was added above the separation tank to remove the oversized particles (clay lumps and rocks) not broken up in the mixer.

4. A single separation vessel was used which had twice the capacity of the separators in the Edmonton plant. The bitumen froth was again removed with a rotating steel drum. The major difference in this vessel was the removal of a single sand, water mixture from the bottom boot section in place of the bucket line.

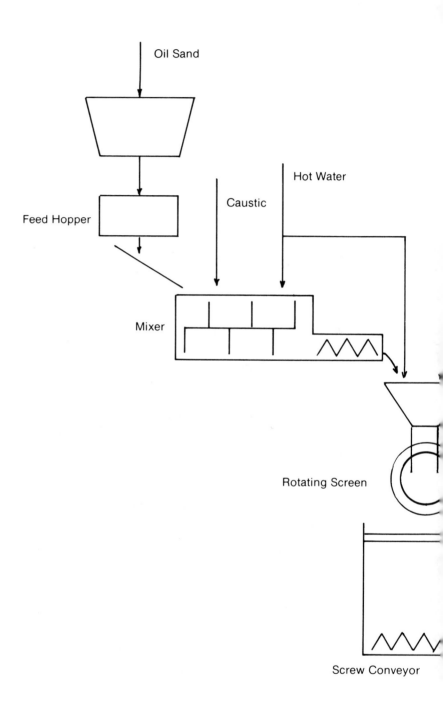

Oil Sand

Feed Hopper

Caustic

Hot Water

Mixer

Rotating Screen

Screw Conveyor

Figure 12. Second Alberta Research Council Plant (1929),
Waterways

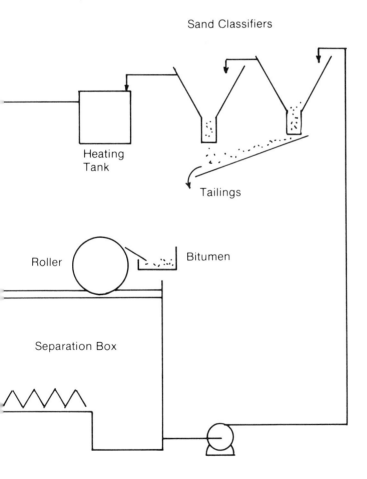

Sand Classifiers

Heating
Tank

Tailings

Roller

Bitumen

Separation Box

5. The sand and clay solids were removed from the separator bottom stream by two classifiers.* The first classifier withdrew the high density sand-water mixture continuously from the bottom. The second classifier removed the finer clay solids intermittently as they settled to the bottom of the cone.

6. The Waterways plant also contained secondary treatment of the bitumen froth to remove excess water and solids. This processing consisted of a steam jacketed mixing vessel, a steam heated settling tank, a steam heated dehydrator, and a final settling tank. The secondary treatment consisted of heating the bitumen to a higher temperature (over 110°C) to increase the density difference between the bitumen and the sand-water mixture. The bulk of the sand water was drawn off the first settling vessel. Next, the heat was maintained in a shallow evaporating vessel to vapourize the remaining water. The final long term settling produced a bitumen with about 5 percent solids and 1 percent water. Although this method has been replaced by dilution centrifuging today, it was the first attempt to produce a highly pure bitumen product.

Many operating procedures were tried at the Waterways plant to solve the problems that existed in the first Edmonton plant and the new problems that arose with the new equipment. The removal of the clay fines was aided by the addition of calcium chloride salt to the water to coagulate the clay into larger particles for easier removal. Also and significantly, concentrated salt brine (20 percent sodium chloride) was tried with good results. The brine increased the density of the middlings stream and allowed more bitumen to float to the surface while not impeding the settling of the sand. This could not be used in today's plants because of pollution problems created by salt water tailings ponds.

The Waterways oil sand had a natural pH of under 5 and required large amounts of silicate of soda to get high bitumen recoveries. Oil sand brought in from the International Bitumen Company at Bitumount was processed at Waterways and was found to separate much easier even without a silicate of soda reagent. The

* Classifiers are cone shaped vessels used commonly in mineral processing as concentrating devices.

pH of the Bitumount oil sand was discovered to be much higher than that of Waterways (7.8). The pH of the Waterways sand was then raised by the addition of sodium hydroxide (caustic) and bitumen recovery increased while the mineral content of the bitumen was reduced dramatically. The addition of the sodium hydroxide for pH adjustment is present in today's plants.

After the extensive runs conducted at Waterways, Clark and Pasternack concluded that the commercial extraction of bitumen with the hot water process was feasible. They foresaw the present-day problems of the clay fines. They also realized the design of the initial mixing vessel could be improved. They did not foresee the scale of power generation needed in a large plant. (The Waterways plant used a wood-fired boiler.) They did not even anticipate pollution problems. Tailings from Dunvegan were fed directly into the Clearwater River.

Today's commercial plants differ only in two major respects from the Clark plant:

1. The new plants use flotation cells to treat the middlings-water stream and increase bitumen recovery.

2. The bitumen froth is treated with dilution centrifuging rather than the evaporation principle used at Waterways.

In spite of these differences, it can be said that the process used in today's commercial plants began with the Waterways plant in 1929.

B. International Bitumen Company
The first oil sands venture to make a commercial product by surface extraction of bitumen was that of R.C. Fitzsimmons and the International Bitumen Company at Bitumount. His failed experiments with drilling to recover bitumen in the 1920s drew him to the conclusion that mining and surface extraction would be the best commercial process. Fitzsimmons examined the existing extraction technology and built his initial plant using the hot water process. His plant was primitive and consisted mainly of scavenged equipment but it worked sufficiently well to produce a roofing tar used in Edmonton. He later produced the first products refined adequately to be used as commercial fuels.

A diagram of Fitzsimmons's plant is shown in Figure 13. The oil sand was mined with a small dragline and dropped into a small hopper. From there it was crushed by a set of toothed rollers and flushed with hot water into a pug mill. A pug mill is a device for mixing thick slurries such as mortar or baking dough by means of "rakes" similar to the blades on a push-type lawn mower. More steam and hot water were added after the initial slurry left the pug mill and the mixture was placed in a long, shallow separation cell. This type of separator is an oil-field device used to separate oil-water mixtures by skimming the oil off the top. The bitumen was skimmed off the top of this vessel and dropped through a screen to remove any large lumps of clay or wood. The tailings were removed by an auger at the bottom of this separator along with a high percentage of the hot water. The bitumen was then stored in two settling tanks to allow a longer period for additional separation of water. Finally, the bitumen was pumped to a storage tank.

The plant used no reagents such as silicate of soda or sodium hydroxide. The bitumen product from the second settling vessel was reported to be of high quality with very little water remaining in it. The high water pH, low clay content and high bitumen percentage make the Bitumount sands among the best in the Athabasca deposit. This greatly aided Fitzsimmons's separation. Also, the extraction plant was run intermittently on a batch basis and the bitumen often sat in the two settling tanks for days. This allowed the water to settle out over this period and the separated water to be drained away before the bitumen was sold as roofing tar or later refined.

Fitzsimmons's plant had several major operating problems. Since the screen was not placed prior to the pug mill, small rocks entered the pug mill. Thus, the rakes were either jammed or worn out. Skimming the bitumen from the separation tank did not work well because of bad design. Most of the hot water was lost with the tailings and made for poor energy efficiency. Finally, while the bitumen product was water free, it contained a high percentage of solids (sand). This caused continuous problems with pump and valve wear in the refinery section.

The design recovery rate was 700 barrels of bitumen per day but a "good" day produced about 250 barrels. There were no technical advances in Fitzsimmons's plant. The plant's main

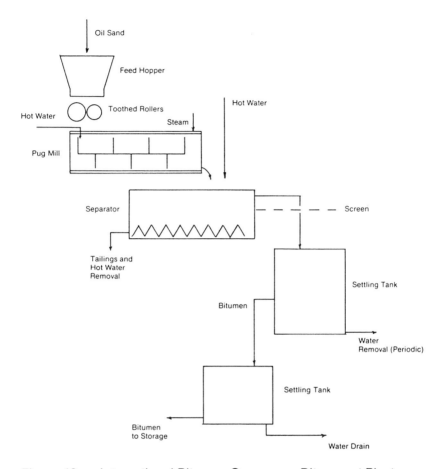

Figure 13.   International Bitumen Company—Bitumount Plant

contribution came from its survival as a commercial venture. Products were recovered from a "jury-rig" operation built on the cheap. Fitzsimmons's remarkable efforts (a testament to employees like Adkins) encouraged attempts by other companies to recover commercial products.[20]

C. Abasand

The second commercial attempt at bitumen recovery and processing was that of Abasand Oils Ltd. The major portion of the process was designed by J.M. McClave from his original work and patented process developed in the mid 1920s. The financing

201

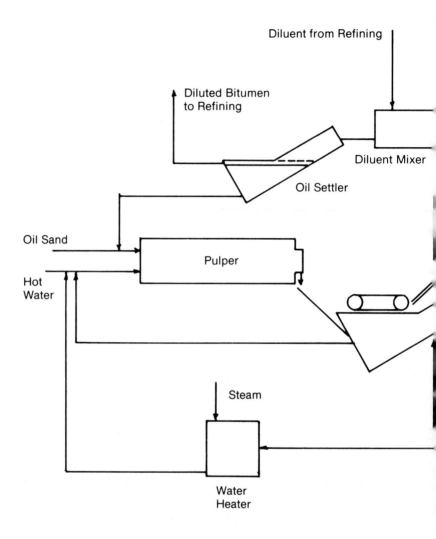

Diluent from Refining

Diluted Bitumen
to Refining

Diluent Mixer

Oil Settler

Oil Sand

Pulper

Hot
Water

Steam

Water
Heater

was arranged by Max Ball from experience he gained in the United States oil industry in the prior two decades. The plant was located on the Horse River two miles south of Fort McMurray.

The process was a variation of hot water extraction as tested by Clark at his Waterways pilot plant. It is shown in Figure 14. The extraction of bitumen from oil sand consisted of the following steps:

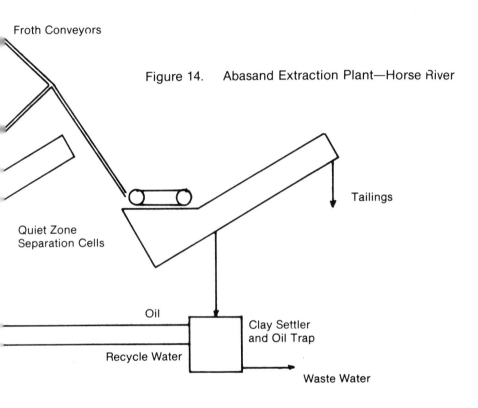

Froth Conveyors

Figure 14.   Abasand Extraction Plant—Horse River

Tailings

Quiet Zone
Separation Cells

Oil

Clay Settler
and Oil Trap

Recycle Water

Waste Water

1. The oil sand, hot water (slightly cooler than Clark's process),
    and alkaline salt (soda ash) were mixed in a pulper. The
    pulper was a large rotating vessel with internal shelves
    similar to a large clothes drier. It was the first initial mixing
    device where the entire vessel rotated instead of having
    mixing blades inside the vessel to do the agitating. This
    design is present in today's plants; it is called a conditioning
    drum by Suncor, a tumbler by Syncrude.

203

2. The pulper slurry was sent to the primary separation vessel which McClave called his Quiet Zone Flotation cell. The flotation cell was essentially an ore classifier (a device for concentrating ore materials from liquid slurries), with steam jacketed heating to maintain slurry temperature. The vessel was rectangular with a sloping bottom. The bitumen floated to the surface and was removed by a large roller which revolved slowly in the bitumen layer. The bitumen adhered to the roller and then was removed with fixed scrapers. The sand sank to the bottom of the classifier and was raked up the inclined bottom of the vessel. The sand was then fed into a second, similar classifier to remove more of the bitumen with further settling and agitation. The bitumen was removed from the second classifier by a similar roller. The sand raked from the second classifier was discharged as solid tailings (80 percent sand) and stockpiled. The warm water middlings was pumped from the middle of the two flotation cells, reheated and reused in the pulper.

3. The bitumen scraped off the rollers was then mixed with an equal amount of naphtha diluent in a mixer. The naphtha lowered the viscosity and density of the bitumen and allowed the remaining sand and water to settle out. This gravity separation was accomplished in a deep rectangular settling vessel. The sand and water dropped to the rounded vessel bottom where it was augered out and fed back to the pulper to eventually be discharged from the second flotation cell. The water was withdrawn above the sand, reheated and again added to the pulper. The diluted bitumen overflowed a weir at the top of the settler and was sent to the refinery section for processing and the recovery of the diluent naphtha.

4. The problem found in Clark's Waterways plant of fine clays building up in the recirculated water again occurred. An additional water settling vessel was added to process the middlings water and this allowed the fine clay particles to settle out.

The first Abasand plant never did achieve sustained design capacity of 3,000 barrels of bitumen per day but did manage to average 400 barrels of oil per day in its last summer of operation in 1941. In fact, the refinery section of the plant was unable to keep up with the extraction process and the hot water extraction plant had to be shut down periodically to let the refinery catch up. The final diluted bitumen product was of much higher quality than any so far produced and contained only slightly more water than found in today's plants (1 percent). The solids content of about 0.5 percent was much higher than found in today's plants.[21]

The first plant was destroyed by fire in November 1941 but the original Abasand plant did advance oil sands extraction technology in three areas. First, as mentioned before, the pulper design was a great improvement over previous methods and is essentially used today. Second, the quiet zone flotation cells have been greatly modified in today's plants and now exist for the secondary treatment of the middlings stream. Finally, the diluent treatment of the raw bitumen followed by gravity separation is very similar to today's bitumen treating. The amount of diluent added was about the same as is used today: Abasand used 1 kilogram diluent per kilogram bitumen, Syncrude .9 kilogram diluent per kilogram bitumen. The diluent bitumen is centrifuged today to accelerate the settling of the sand and water instead of the gravity settler.

After the initial plant was destroyed, Abasand started to rebuild the plant with no major changes in the initial extraction design. Money and manpower shortages caused the company to abandon its operation but the Canadian government took over the plant in 1943. They contracted the General Engineering Company of Toronto (a mineral processing design firm) to expand McClave's design to produce 4,000 barrels of bitumen a day. They experimented with treatment of the bitumen-hot slurry with conventional mineral flotation cells (air injected). They also adapted the "cold water process" developed by Djingheu- zian as described earlier (Figure 5) to McClave's equipment. They added diluent at the pulper with 25°C cooler water. The bitumen recovery was very high (92 percent) but the diluent losses also were very high in the sand tailings (40 percent loss). More flotation cells were added to increase the diluent recovery.

Because the pulper did not contain the balls suggested in the Djingheuzian design, the bitumen-cold water-diluent slurry produced when cold or weathered oil sand was treated was lumpy and did not allow for good separation. Also, the lumps caused excessive wear on the blades of the pulper.

In August 1945 the second plant was also burnt down. The hybrid design of the second plant - i.e. using hot water equipment with cold water techniques - was not proven. The cold water technique was essentially abandoned until energy conservation concerns revived it in the late 1970s.[22]

D. Government of Alberta Plant - Bitumount
The last large scale pilot designed and built specifically to prove the economic feasibility of oil sand extraction was constructed beside the original Fitzsimmons plant at Bitumount in 1946 and 1947. The plant ran periodically for two seasons beginning in the later summer of 1948. It sustained the technical feasibility of the hot water extraction method but did not prove the economic viability of the process compared to the production of conventional crude being discovered at the same time.

The plant was built by the Born Engineering Co. of Tulsa, Oklahoma. The design was substantially different from the original Alberta Research Council plant at Waterways but was still based on Clark's original hot water method. The process is shown in Figure 15 and consisted of the following steps:

1. The mined oil sand was fed into a large hopper covered with a coarse screen to reject oversize material.

2. A steam-jacketed screw conveyor moved the oil sand from the bottom of the feed hopper. Hot water was added in the screw conveyor to help dilute the oil sand and create a slurry.

3. The thinned oil sand passed through a rotating screen to remove any remaining rocks or clay lumps.

4. The diluted oil sand then entered a steam-jacketed pug mill for further mixing and heating. After the pug mill, more hot water was added to wash the bitumen into the separation cell. The water was forced into the slurry at high velocity to cause very turbulent flow. If the air added by the water turbulence was not sufficient, supplemental air could be added to the water to help the flotation process.

5. The separation cell receiving the bitumen-water slurry was a large rectangular tank. The sand settled to the bottom and

was removed with a screw conveyor as high solid content tailings. The middling stream was pumped to another large vertical tank to allow the bulk of the fine clay solids and silt to settle out and be removed. This water was then reheated with steam and recirculated. The bitumen that floated to the top of the separation cell was skimmed off the top. It contained up to 35 percent water and 5 percent solids and was sent to a second separation cell for further treating.

6. The second separation vessel was similar to the first except that it had a steeply sloped bottom. The added retention time allowed more of the water and solids to settle out of the bitumen, however, the solids were covered with oil and were not sent to tailings but recycled back to the pug mill.

7. The bitumen skimmed from the second separator was next diluted with a kerosene diluent and placed in an agitated mixer. The mixing and additional settling time allowed 70 to 75 percent of the remaining water and fine sand to drop out of the diluted bitumen. Revolving rakes swept the water-solid mixture out of this vessel to be disposed of as additional tailings.

8. The diluted bitumen overflow still contained excess water and was removed by retorting. This process consisted of heating the diluted bitumen to a temperature high enough to vapourize the remaining water but not high enough to vapourize much of the bitumen and diluent. The water was then condensed and any vapourized oil recovered with a conventional oil separator. The retorted "dry bitumen" was sent to storage for later refining.

The plant was designed to recover 350 barrels of bitumen per day and in fact consistently produced 500 barrels per day. The individual vessel design was the most rigorous of any plant yet built and this contributed to the overcapacity production. However, the design concept had several flaws and was very far from today's commercial extraction method.

First, the initial mixing of oil sand and water by screw conveyors and pug mill produced a good slurry but the closed design did not permit enough air to be entrained with the bitumen. This prevented good bitumen flotation and efficient recovery. Initial mixing with a tumbler introduces much more air and hence

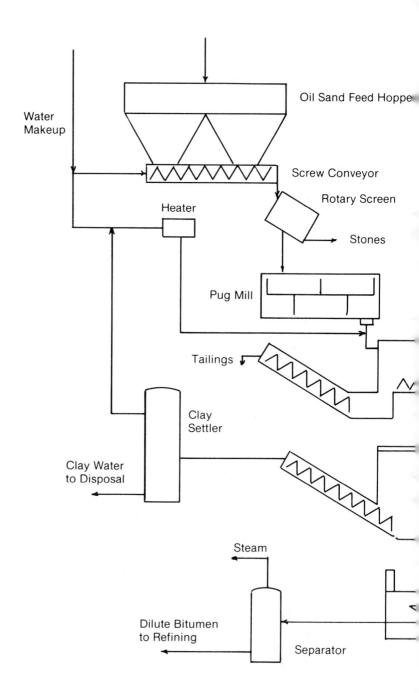

Oil Sand Feed Hopper

Water
Makeup

Screw Conveyor

Rotary Screen

Heater

Stones

Pug Mill

Tailings

Clay
Settler

Clay Water
to Disposal

Steam

Dilute Bitumen
to Refining

Separator

Figure 15.   Alberta Government Plant—Bitumount

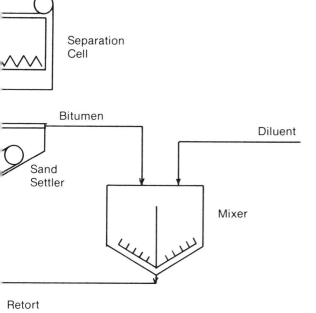

Separation Cell

Bitumen

Diluent

Sand Settler

Mixer

Retort

better flotation. Second, the tailings removal as a highly solid stream would present difficulties in scale-up to larger plants and is not used because of these solid handling difficulties. Also, the removal of clay solids by gravity settling required huge tanks in a large scale operation and would not be practical. The second separation vessel for removal of solids and water from the initial bitumen proved very ineffective and could have been left out. Finally, the secondary treatment of the bitumen by kerosene dilution did not perform satisfactorily. Kerosene was not as effective at lowering the viscosity and density as a lighter naphtha (used in today's plants). However, the subsequent retorting process prevented the use of a lighter naphtha because of the excess vapour losses with the more volatile diluent. In other words, by using the retorting process, a heavy diluent had to be used instead of a lighter one that could have prevented the need for retorting in the first place.[23]

The design of the second plant at Bitumount bore little resemblance to today's commercial plants. The design faults were discovered very quickly. The modified design suggested by Blair in 1951 is much closer to today's processes.[24] However, the plant did prove that, if equipment was designed and built specifically for bitumen extraction based on established engineering principles and using bitumen fluid properties, it could successfully extract bitumen over an extended period of time.

* * *

All the plants build in the Fort McMurray area since 1929 have contributed to the technology of oil sand extraction. The initial Alberta Research Council plants pioneered the development of the hot water extraction method. The Clearwater plant's operation helped establish the best hot water temperature, the correct water-oil sand ratio and the best operating pH level for efficient extraction. The mechanical design also demonstrated some of the problems uniquely associated with oil sand processing and prevented later designs from making the same errors.

Fitzsimmons's International Bitumen plant demonstrated the

feasibility of extracting oil sand and selling a commercial product. While the technological advances of his plant were limited mainly to showing what not to do, his tenacity and enthusiasm for the oil sands helped encourage other, more scientific attempts.

Abasand Oils Ltd. improved the hot water extraction process by developing the initial tumbler design, using diluent to treat the bitumen and developing quiet zone separation vessels. If the misfortunes of the two fires and the Leduc discovery had not shut down this operation, larger scale ventures may have developed sooner from this plant's design.

The final Alberta government Bitumount plant proved that, if equipment was designed specifically for the extraction of bitumen, it could run successfully for sustained periods of time and produce a product suitable for further processing.

The plants now producing synthetic crude oil from surface extraction of bitumen will continue to improve extraction technology. The problems associated with the disposal of sand tailings and the large energy requirements of present designs will continue to be investigated by government and private industry. Solutions may come from some of the alternate proposals reported in this Appendix or further changes to the hot water process. In-situ extraction is still in the research stage but plans to exploit this technique are beginning to shift to the commercial exploitation stage.

ABBREVIATIONS USED IN NOTES

APS   Alberta. Provincial Secretary, Provincial Archives of Alberta

CM&S   Consolidated Mining and Smelting, Provincial Archives of Alberta

ERCB   Energy Resources Conservation Board, Provincial Archives of Alberta

HMT   Henry Marshall Tory, Presidents' Papers, University of Alberta Archives

KAC   Karl Adolph Clark, University of Alberta Archives

LAR   Louis A. Romanet, University of Alberta Archives

MM   Mining and Metallurgy, University of Alberta, University of Alberta Archives

PAA   Provincial Archives of Alberta

PAC   Public Archives of Canada

PP   Premiers' Papers, Alberta, Provincial Archives of Alberta

PW   Public Works, Alberta, Provincial Archives of Alberta

PWM   Public Works, Ministers' Files, Alberta, Provincial Archives of Alberta

RCA   Research Council of Alberta, University of Alberta Archives

RCF   Robert C. Fitzsimmons, Provincial Archives of Alberta

RCW   Robert Charles Wallace, Presidents' Papers, University of Alberta Archives

RG   Record Group, Public Archives of Canada

TD   Transportation Department, Provincial Archives of Alberta

TDM   Transportation Department, Ministers' Files, Provincial Archives of Alberta

UAA   University of Alberta Archives

Notes

INTRODUCTION:

1. Alberta Energy and Natural Resources, "Alberta Oil Sands Facts and Figures," Energy and Natural Resources *Report No. 110* (1979), vii-viii; Donald Towson, "Tar Sands," *Encyclopedia of Chemical Technology*, 3rd ed. (New York: John Wiley and Sons, 1978); Max W. Ball, *This Fascinating Oil Business* (Indianapolis and New York: The Bobbs Merrill Company, 1940), contains clear discussion of the physics and chemistry of petroleum as well as of the Athabasca tar sands; Larry Pratt, *The Tar Sands: Syncrude and the Politics of Oil* (Edmonton: Hurtig Publishers, 1976), 33-34, also has a useful description of the resource.

2. Alberta Energy Resources Conservation Board, "Crude Bitumen Reserves ... 1980," ERCB *Report* 81-38 (1981); Alberta Energy and Natural Resources, "Alberta Oil Sands Facts and Figures," 5; G.J. Demaison, "Tar Sands and Supergiant Oil Fields," Canadian Institute of Mining, 17 Annual (1977), 9-11; P.M. Dranchuk, "Exploitation of the Oil Sands of Alberta," *Proceedings, Ninth Annual Conference*, Canadian Technical Asphalt Association 19 (1974): 171-86; R.M. Procter, Gordon C. Taylor, John A. Wade, "Oil and Natural Gas Resources of Canada 1983," GSC *Paper 83-31* (Ottawa: Geological Survey of Canada, 1983).

3. Alberta Energy and Natural Resources, "Alberta Oil Sands Facts and Figures," 5, 11; Canadian Petroleum Association, *Statistical Handbook* (Calgary: Canadian Petroleum Association, 1980), table II-I.

4. Alberta Energy and Natural Resources, "Alberta Oil Sands Facts and Figures," 5, 11. On the variations in estimates of reserves see John M. Blair, *The Control of Oil* (New York: Vintage Trade Books, 1978), 4-23.

5. Proctor, Taylor, Wade, "Oil and Natural Gas Resources of Canada 1983."

6. Alberta, Department of Provincial Treasurer, *Government Estimates*, 1985-86 (Edmonton: 1985), 150-51; Alberta,

Department of Provincial Treasurer, *Public Accounts of Alberta*, 1980-81, (Edmonton: n.d.) statements 5,50, 5,86; Alberta Oil Sands Technology and Research Authority (AOSTRA), *Fifth Annual Report and Five Year Review* (Edmonton: Alberta Oil Sands Research Authority, 1980); *Oilweek* 36, no. 6, (11 March 1985): 15.

I. THE DOMINION GOVERNMENT

1. John W. Chalmers, "A Century of the Fur Trade," in *The Land of Peter Pond*, edited by John W. Chalmers (Edmonton: Boreal Institute of Northern Studies, University of Alberta, Occasional Publication no. 12, 1974), 45-62; C.S. MacKinnon, "Some Logistics of Portage La Loche," *Prairie Forum* 5, no. 1 (1980): 51-65; Pond left no writings on the Athabasca country: see Charles M. Gates's collection of Pond's trip accounts, *Five Fur Traders of the Northwest* (Minneapolis: University of Minnesota Press, 1933).

2. W. Kaye Lamb, ed., *Journals and Letters of Alexander Mackenzie* Hakluyt Society, Extra Series, no. 4 (Toronto: Macmillan of Canada, 1970), 128-29. None of Mackenzie's published letters comment on the significance of the patching uses of bitumen.

3. John Richardson, *Arctic Searching Expedition: A Journal of a Boat-Voyage Through Rupert's Land and the Arctic Sea*, 1 (London: Longman, Brown, Green and Longmans, 1851), 124-27.

4. The impact of the Canadian acquisition is surveyed in several important books: W.L. Morton, *The Critical Years: The Union of British North America, 1857-1873* (Toronto: McClelland and Stewart, 1964), 233-44, 245ff; Peter B. Waite, *Canada 1871-1896: Arduous Destiny* (Toronto: McClelland and Stewart, 1971); G.F.G. Stanley, *The Birth of Western Canada: A History of the Riel Rebellion* (London: Longmans, Green, 1936; Toronto: University of Toronto Press, 1961).

5. See Morris Zaslow's indispensable works on the northern frontier of Canada: *Reading the Rocks: The Story of the Geological Survey of Canada, 1842-1972* (Toronto: Macmillan of

Canada in association with the Department of Energy, Mines, and Natural Resources, and Information Canada, 1975), and *The Opening of the Canadian North, 1870-1914* (Toronto: McClelland and Stewart, 1971), 9-10, 80-86. The rich storehouse of Canadian attitudes toward the development of the West, and the impact of the West on those Canadians active in its takeover, are discussed in Douglas R. Owram, *The Promise of Eden: The Canadian Expansionist Movement and the Idea of the West, 1857-1900* (Toronto: University of Toronto Press, 1980), especially Chapters 5 and 7.

6. Zaslow, *Reading the Rocks*, 110, 234.

7. Macoun is examined in Zaslow, *Reading the Rocks*, 132-33, and Owram, *The Promise of Eden*, 152ff; Macoun's work on the Athabasca is found in his "Geological and Topological Notes, by Professor Macoun, on the Lower Peace and Athabasca Rivers," Geological Survey of Canada, *Report of Progress for* 1874-6 (Montreal: by Authority of Parliament, 1877), App. 1: 87-95.

8. Macoun, "Geological and Topographical notes," 92.

9. Ibid., 93.

10. Information on Bell is to be found in Zaslow, *Reading the Rocks*.

11. Robert Bell, "Report on Part of the Athabasca River," in Geological Survey of Canada, *Report of Progress for* 1882-84 (Montreal: Dawson Brothers, 1884), 37cc; 21cc-22cc.

12. Ibid., 24cc.

13. Ibid., 32cc-33cc; Geological accounts of the Athabasca are summarized by the geographer W.C. Wonders, "Wood and Water, Land and Oil," in *The Land of Peter Pond*, 19-20; Wonders notes the Indians' use of bitumen as does James Parker, "The Long Technological Search," in *The Land of Peter Pond*, 110.

14. Bell, "Report," 33cc-35cc, quotation on 34cc; G.C. Hoffman is mentioned in Geological Survey of Canada, *Report of Progress for* 1880-82 (Montreal: Dawson Brothers, 1883), 34.

215

15. Bell, "Report," 34cc.

16. Ibid.

17. On the Senate Report and the general estimate of the area's potential see Zaslow, *The Opening of the Canadian North*, 77-79.

18. Robert G. McConnell quoted in Elfric Drew Ingall, "Division of Mineral Statistics and Mines Annual Report for 1890," Geological Survey of Canada, *Annual Report for 1890-1 5*, Part 2 (Ottawa: Queen's Printer, 1893), 144s-147s. McConnell's estimate is rendered as tonnage in George Harcourt, "Economic Resources of Alberta," in *Canada and its Provinces. A History of the Canadian People and Their Institutions By One Hundred Associates* 20, edited by A. Shortt and A.G. Doughty, Edinburgh edition (Toronto: Edinburgh University Press for the Publishers Association of Canada, 1914), 598.

19. McConnell quoted in Ingall, "Division of Mineral Statistics and Mines," 146s-147s.

20. George M. Dawson, "Summary Report," Geological Survey of Canada *Annual Report* 1894 7 (Ottawa: Queen's Printer, 1896), 6A-14A; David H. Breen, ed., *William Stewart Herron: Father of the Petroleum Industry in Alberta* (Calgary: Alberta Records Publication Board, 1984), Appendix 1.

21. Dawson, "Summary Report," (1894), 12A-14A; George M. Dawson, "Summary Report," Geological Survey of Canada, *Annual Report* 1897 10 (Ottawa: Queen's Printer, 1899). 18A-27A.

22. Dawson, "Summary Report," (1897), 18A-27A; See also Harcourt, "Economic Resources of Alberta," 598; Zaslow, *Reading the Rocks*, 234; George de Mille, *Oil in Canada West: The Early Years* (Calgary: Northwest Printing and Lithographing, 1970), 39-43.

23. George M. Dawson, "Summary Report," Geological Survey of Canada *Annual Report* 1898 11 (Ottawa: Queen's Printer, 1901), 28A-36A.

24. Dawson, "Summary Report," (1898), 31A; Zaslow, *Reading the Rocks*, 234.

25. The two documented and reliable discussions of Hammerstein are by Zaslow, and by Parker and Tingley: Zaslow, *The Opening of the Canadian North*, 210-12; J.M. Parker and K.W. Tingley, "History of the Athabasca Oil Sands Region 1890 to 1960s," 1, Alberta Oil Sands Environmental Research Programme, Project HS 10.2, 1980, 234.

26. Zaslow, *Opening the Canadian North*; David H. Breen, "Anglo-American Rivalry and the Evolution of the Canadian Petroleum Industry to 1930," *Canadian Historical Review* 62 (Summer 1981): 283-303.

27. Zaslow, *Reading the Rocks*, 233-34; Sidney C. Ells, "Recollections of the Development of the Athabasca Oil Sands," Canada, Mines Branch *Information Circular* No. 139, (Ottawa: Queen's Printer, 1962), 1ff. Ells's account is *not* reliable outside of other corroborating evidence.

28. S.C. Ells, "Preliminary Report on bituminous sands," Canada, Mines Branch *Report* No. 281 (Ottawa: King's Printer, 1914); S.C. Ells, "Recollections," vii-ix, preface by W.H. Norrish, 1-13.

29. E. Haanel to R.W. Brock, 10 March 1914; R.G. McConnell to E. Haanel, 24 March 1915; E. Haanel to S.C. Ells, 12 and 26 August 1915; all in PAC, Canada, Mineral Resources Branch, RG 87/30/262-2; See also Ells, "Recollections," 27-30.

30. E. Haanel to R.G. McConnell, 4 August 1917, UAA, RCA, 80/1/2/4-3; R.G. McConnell to H.M. Tory, UAA, HMT, 906-5a-2.

31. The copy used for this study is in the University of Alberta Archives, Research Council of Alberta collection. It is one of two known copies. See RCA, 80/1/2/4-2, 2 unpaginated. The other copy is also at the University of Alberta Archives, Mining and Metallurgy records, MM, 69/76/1.

32. Ells, "Notes," prefatory note, vol. 1. Again, note that the volumes are unpaginated.

33. Ells, "Notes," vol. I; appendices 8 and 18 in vol. II.

34. Ells, "Notes," vol. II; appendices 8 and 17.

35. Ells, "Notes," vol. II; appendix I.

36. Ells, "Notes," vol. II "General Conclusions."

37. n.p.n.d. (Attributed by Eugene Haanel to Karl Clark and Joseph Keele and dated by Haanel as August 1917): "Review of the Report Sumitted by S.C. Ells ...." RCA, 80/1/2/4-3, especially 32; see E. Haanel to R.G. McConnell, 4 August 1917, RCA.

38. E. Haanel to R.G. McConnell, 4 August 1917, RCA, 80/1/2/4-3.

39. G. Parker, "Alberta Bituminous Sands for Rural Roads," Canada, Mines Branch Report for 1918, *Sessional Paper No. 26a*, 1920.

40. S.C. Ells to A. Lehmann, 17 March 1917; John Allan to H.M. Tory, 3 August 1920; both in HMT, 906-5a-2; Edgar Stansfield to H.M. Tory, 20 July 1924, RCA, 80/1/3-3.

41. S.C. Ells to Herbert Greenfield, 22 and 23 October 1924, PAA, PP, 174; Ells's 1917 work was in fact revised and published as the bulk of his 1926 survey of the oil sands: see S.C. Ells, "Bituminous Sands of Northern Alberta: Occurrence and Economic Possibilities," Canada, Mines Branch *Report No.* 632 (Ottawa: King's Printer, 1926), especially 85-113.

II: A PROVINCIAL INITIATIVE

1. Eugene Haanel to R.G. McConnell, 4 August 1917 "confidential," enclosed report, Karl Clark and Joseph Keele evaluation, PAA, RCA, 80/1/2/4-3; K.A. Clark to Eugene Haanel, 5 August 1920, PAC, Canada. Mineral Resources Branch, RD87/31/462.128.

2. Maureen Riddell Aytenfisu, "The University of Alberta: Objectives, Structure and Role in the Community, 1908-1928" (Master's thesis. University of Alberta, 1982), 54-79, 264-74. See also her useful essay on the creation of the Research Council: Maureen Riddell, "Origins and Evolution of the Scientific and Industrial Research Council of Alberta, 1919-1930" (Bachelor's Honours essay, Univer-

sity of Alberta, Department of History, 1977). Cf. Douglas R. Owram, "The Economic Development of Western Canada: An Historical Overview," Economic Council of Canada, *Discussion Paper No.* 219 (1982), which repeats the traditional view that virtually no concern with economic diversification occurred until after the Second World War. Other important examples of this view include the important and generally convincing interpretation of Larry Pratt and John Richards. Again, Pratt and Richards date diversification efforts in Alberta and Saskatchewan to the post-depression period. They are correct in noting that until the 1940s the primary emphasis was on agriculture and that the primary agent of development was the Dominion government. John Richards and Larry Pratt, *Prairie Capitalism: Power and Influence in the New West* (Toronto: McClelland and Stewart, 1979), 9-12. The fact that the beginning of provincially-led diversification occurred prior to the transfer of natural resource jurisdiction should be noted by historians; they may prove instructive to us if they are ever studied definitively.

3. Tory to Frank Adams, 13 January 1920, HMT, 902-3c; Tory to J.L. Coté, 21 January 1920, HMT, 1104-2; Tory to A.B. Macallum, 22, 25, 26 May 1920, Macallum to Tory, 13 March 1920, both in HMT, 906-5a. On the larger support for research from Alberta business and the University's response, see Aytenfisu, "University of Alberta ... 1908-28," 270-72.

4. H.M. Tory to H. Greenfield, 28 February 1924, HMT, 1104; cf. A.B. Macallum to H.M. Tory, 16 June 1920, HMT, 906-5a.

5. Minutes, Committee on Industrial Research, 25 May 1919, 27 June 1919, HMT, 1104-2; "Agreement Between University of Alberta and Provincial Government relating to ... Industrial Research," 20 January 1921, UAA, RCA, 80/1/2-1; on J.L. Coté, see M. Riddell, "Origins and Evolution," 25, 32. On Karl Clark's career, a number of sources provide basic information. The detailed study of his career that his achievements and records merit awaits the full availability of his private as well as public papers. Currently Clark's scientific work can be well studied through Alberta Research Council and University Department of Mining and

Metallurgy records at the University of Alberta Archives, and through Karl Clark's scientific diaries available at the Provincial Archives of Alberta. But until the University of Alberta Archives gains permission to open his private papers, the full story of his research and of much else about engineering and business development in Alberta will not be complete.

6. W.M. Edwards to H.M. Tory, 31 December 1912; Adolph Lehman to Sidney Ells, 27 March 1917; H.M. Tory to S.C. Ells, 7 April 1917; British Comptroller of Munitions to H.M. Tory, 12 April 1916; all in UAA, HMT, 906-5a-2. Tory's interest is noted in J.M. Parker and K.W. Tingley, "History of the Athabasca Oil Sands Region 1890-1960s," Alberta Oil Sands Environmental Research Program Project HS 10.2, 1980, 121-22.

7. A.B. Macallum to H.M. Tory, 13 March 1920, HMT, 906-5a; A. Lehman to H.M. Tory, 22 March 1920, HMT, 906-5a; H.M. Tory to A.B. Macallum, 23 March 1920, HMT, 906-5a.

8. August W. Giebelhaus, *Business and Government in the Oil Industry; A Case Study of Sun Oil, 1876-1945* (Greenwich, Conn.: JAI Press, 1980), 198 ff.; John M. Blair, *The Control of Oil* (New York: Vintage Trade Books, 1978), ch. 6 & 7.

9. See David H. Breen ed., *William Stewart Herron: Father of the Petroleum Industry in Alberta.* (Calgary: Alberta Records Publication Board, 1984); and George de Mille, *Oil in Canada West: The Early Years* (Calgary: Northwest Printing and Lithographing, 1970).

10. K.A. Clark to H.M. Tory, 19 November 1921, 8 August 1921, K.A. Clark, "Tar Sands Investigation," 18 November 1921, both in RCA, 80/1/2/4-1.

11. K.A. Clark to H.M. Tory, 19 November 1921, RCA, 80/1/2/4-1; Research Council of Alberta, "Estimated Expenditures for 1921," RCA, 80/1/1/3-1.

12. E. Stansfield to H.M. Tory, 6 April 1922, RCA, 80/1/3-3; K.A. Clark to H.M. Tory, 22 September 1922, RCA, 80/1/2/4-1.

13. H.M. Tory to Charles Stewart, 6 September 1922, "personal and private," HMT, 906-5a-2.

14. Research Council of Alberta, *Second Annual Report* (Edmonton: Research Council of Alberta, 1922), 43-59; Research Council of Alberta, *Third Annual Report* (Edmonton: Research Council of Alberta, 1923), 42ff.

15. K.A. Clark and S.M. Blair, "The Bituminous Sands of Alberta, 1, Occurrences." Research Council of Alberta *Report No.* 18 (Edmonton: Research Council of Alberta, 1927).

16. Ibid., 9-11, especially 10.

17. Ibid., 63-64, 68-69.

18. Ibid., 65-68, 73-74.

19. Clark and Blair, "The Bituminous Sands of Alberta, 2, Separation," Research Council of Alberta *Report No.* 18 (Edmonton: Research Council of Alberta, 1928), 1-2.

20. Ibid., 3.

21. Ibid., 6-7.

22. Ibid., 8-15, especially 12; see also K.A. Clark to W.H. Brooks, 12 December 1923, HMT, 906-5a-2.

23. Clark and Blair, "The Bituminous Sands of Alberta, 2, Separation," 16-19, especially 18.

24. Ibid., 19-22.

25. Ibid., 24-26.

26. Ibid., 28-32.

27. Clark and Blair, "The Bituminous Sands of Alberta, 3, Utilization," Research Council of Alberta *Report No.* 18 (Edmonton: Research Council of Alberta, 1929), 1-2, 31-32.

28. Research Council of Alberta, *Seventh Annual Report* (Edmonton: Research Council of Alberta, 1926), 41-43; Research Council of Alberta, *Eighth Annual Report* (Edmonton: Research Council of Alberta, 1927), 44-48; Research Council of Alberta, *Ninth Annual Report* (Edmonton: Research Council of Alberta, 1928), 12-13.

29. K. Clark, "Report of Observations," RCA, 80/1/2/4-1; K. Clark, Diary, mss. 1926, entries for 16-19 August 1926, 23

August 1926, UAA, KAC, Box 1. Earl Smith, "Lab. Report on Samples of Oil Extracted from Oil Sands of Alberta," 3 December 1926, PAC, Earl Smith Papers.

30. K. Clark, Diary, mss. 1926, entries for 16-19 August, 23 August 1926, KAC, Box 1.

31. K. Clark, Diary, mss. 1926, entries for 30 August, 2 September 1926.

32. K. Clark in Research Council of Alberta, *Third Annual Report*, 48-58, especially 52-54. This summarizes his view of a number of both private and public works projects on paving which used oil sands crudely mixed with gravel or simply laid down after heating and rolling.

33. Clark in Research Council of Alberta, *Eighth Annual Report*, 42-48; and Research Council of Alberta, *Ninth Annual Report*, 12-13.

34. S.C. Ells, "Use of Alberta Bituminous Sands for Surfacing Roads," Canada, Department of Mines, *Mines Branch Report No. 684* (Ottawa: King's Printer, 1927).

35. Clark and Blair, "The Bituminous Sands of Alberta, 3, Utilization," 6-9, 14, 31.

36. Quotation: K. Clark, "Progress Report," 30 November 1927, RCA, 80/1/2/3-6; K. Clark, "Memorandum," 2 December 1927, RCA, 80/1/2/4-1.

37. Research Council of Alberta, "Minutes," 28 December 1928, 9 February 1929, RCA, 80/1/3-2; K. Clark, "Memorandum," 26 March 1929. UAA, RCW, 3/2/6/2; J.E. Brownlee to H. Greenfield, 28 December 1928, PAA, PP, 525; R.C. Wallace to Charles Camsell, 1 and 18 April 1929, RCW, 3/2/6/2. The administrative history of the committee may be found in detail at PAC in federal Mineral Resources Branch RG 87/30/462. Copies of the memorandum of agreement dated 13 May 1929 are in both the files of President Wallace of the University and Premier Brownlee: see RCW, 3/2/6/2 and PP, 174. The agreement was printed in the Research Council's *Tenth Annual Report* (Edmonton: Research Council of Alberta, 1929), 59-60.

38. See references cited in note 37.

39. On the joint studies conducted, see one quick summary in R.C. Wallace to J.E. Brownlee, 26 February 1929, 14 May 1929; and Brownlee to Wallace, 17 May 1929; both in RCW, 3/2/6/2/4 and 5. One contemporary observer of the wastage at Turner Valley who explained its causes and dimensions with equanimity was Max Ball, *This Fascinating Oil Business* (Indianapolis and New York: The Bobbs Merrill Company, 1940), 371. Ball noted that the annual gas use of New York was burned off at Turner Valley in the thirties. Earle Gray states that the wasted gas (600 million cubic feet per day) would have served Toronto's daily needs as of 1980. Earle Gray, *Wildcatters: The Story of Pacific Petroleums and Westcoast Transmission* (Toronto: McClelland and Stewart, 1982), 45.

40. Research Council of Alberta, "Appropriations 1923 to 1932-33," RCA, 80/1/1/2-2; Bill 37, Research Council of Alberta Act, text and correspondence in RCA, 80/1/1/2-1; cf. Research Council of Alberta, "Minutes," March 1933, suspending the Council's research programmes, RCA, 80/1/3-2.

41. Public Accounts, Alberta Department of Provincial Treasurer, PAA, Selected Years.

42. K. Clark to R.C. Wallace, 14 May 1929, RCA, 80/1/5-1; K. Clark, "Progress Report," 10 June, 20 July, 6 December 1929, RCW, 3/2/6/2; Research Council of Alberta, "Minutes," 15 November 1930, RCA, 80/1/3-2; Research Council of Alberta, *Tenth Annual Report*, 48-59.

43. K. Clark, Diary mss., entries for 7-31 May, 7 June, 9-22 June, 28 June 1930, KAC, Box 1; K. Clark, "Progress Report," October 1930, RCA, 80/1/2/3-6; Research Council of Alberta, *Eleventh Annual Report* (Edmonton: Research Council of Alberta, 1930), 45-56, especially 50-51.

44. Clark in *Eleventh Annual Report*, 61.

45. Research Council of Alberta, "Appropriations 1923 to 1932-33;" K. Clark in Research Council of Alberta, "Minutes," 10 February 1931; both in RCA, 80/1/3/2; cf. "Minutes," 19 December 1931, RCA, 80/1/32; George Hoadley to R.C. Wallace, 12 August 1930, RCW, 3/2/6/2.

46. Research Council of Alberta, "General Statement," RCA, 80/1/1/3-1; John McLeish to G.G. Ommanney, 22 March 1930, copy in RCW, 3/2/6/2; A.W.G. Wilson, "Memorandum," 12 December 1934, PAC, Geological Survey of Canada, RG 45/34/7.

III: PROMOTIONS AND EXPERIMENTS ON THE ATHABASCA

1. Ells, "Bituminous Sands of Northern Alberta," Canada, Mines Branch *Report* No. 632 (Ottawa: King's Printer, 1926), 21-27; Zaslow, *The Opening of the Canadian North,* 1870-1914 (Toronto: McClelland and Stewart, 1971), 210-12.

2. Note the clear explanation of this aspect of Hammerstein's activity by J.M. Parker and K.W. Tingley, "History of the Athabasca Oil Sands Region 1890-1960s," 1, Alberta Oil Sands Environmental Research Program Project HS 10.2, 1980, 113-15; on Hammerstein's later career see: K. Clark to R.C. Wallace, 7 February 1930; A. Hammerstein to R.C. Wallace, 21 April 1932; R.C. Wallace to A. Hammerstein, 22 April 1932; all in UAA, RCW, 3/2/6/2. See also "Von Hammerstein's Loan with L. Romanet," UAA, LAR, 7/1/2.

3. Ells, "Bituminous Sands of Northern Alberta," (1926), 21-27.

4. Zaslow, *Opening of the Canadian North,* 211-12; see also M. Bladen, "Construction of Railways in Canada, II: 1885-1931," *Contributions to Canadian Economics* 7 (1934): 91.

5. Ells, "Bituminous Sands of Northern Alberta," (1926), 21-27.

6. Ells, "Bituminous Sands of Northern Alberta," (1926), 21-27; Paul von Auerberg, "Report" (on drilling operations), 1 September 1923; H.H. Rowatt to R.C. Fitzsimmons, 28 April 1923; both in PAA, RCF, 11/316; see also George de Mille, *Oil in Canada West: The Early Years* (Calgary: Northwest Printing and Lithographing, 1970), 51-52 and J.M. Parker and K.W. Tingley, "History of the Athabasca Oil Sands Region," 1, 116-18.

7. K. Clark in Research Council of Alberta, *Eighth Annual Report,* (Edmonton: Research Council of Alberta, 1927), 42-48; S.C. Ells, "Uses of Alberta Bituminous Sands for

Surfacing Roads," Canada, Mines Branch *Report No.* 684 (Ottawa: King's Printer, 1927).

8. Ells, "Bituminous Sands of Northern Alberta," (1926), 210; J.M. Douglas to H.M. Tory, 12 March 1921, UAA, HMT, 906-5a-2. A good narrative of Draper's career is by D.J. Comfort, "Tom Draper: Oil Sands Pioneer," *Alberta History* 25, no. 1 (Winter 1977): 25-29.

9. K. Clark to H. Greenfield, 27 October 1922; H. Greenfield to T. Draper, 20 October 1922; T. Draper to H. Greenfield, 17 March 1922; all in PAA, PP, 174. H. Greenfield to T. Draper, 17 April 1924; T. Draper to H. Greenfield, 19 April 1924; K. Clark to H. Greenfield, 24 April 1924; all in PP, 174. R.C. Wallace to T. Draper, 26 November 1930, RCW, 3/2/6/2. A most useful file on Draper's extensive mining efforts from 1925 to at least 1935 has recently been made available in the Provincial Archives of Alberta (as of February 1983) but was not available when this book was being researched. But see Energy Resources Conservation Board collection, ERCB, 6/138. Comfort, "Tom Draper," 27; Canada, Mines Branch, "The Canadian Mineral Industry in 1935," Mines Branch *Report No.* 773 (Ottawa: King's Printer, 1936), 40.

10. Ells, "Bituminous Sands of Northern Alberta," (1926), 210-12; cf. Comfort, "Tom Draper," who claims that the Separation Plant cost $30,000. While Ells was not always reliable, Comfort provides no documentary proof for her estimation. On the Clark-Draper quarrel, see: H.M. Tory to H. Greenfield, 13 June 1923, HMT, 1104-1; K. Clark to H. Greenfield, 19 June 1924, PP, 174. Cf. Clark's diary entries about cordial meetings with Draper later: K. Clark, Diary, mss., entries for 29 October 1929, 7 September 1927, PAC, KAC, Box 1.

11. T. Draper, McMurray Asphaltum, open letter, 10 July 1925, PP, 174; T. Draper to C. Camsell, 27 January 1925; C. Camsell to C. Price-Green, 1 April, 27 May 1925; T. Draper to C. Stewart, 20 November 1925; J. McLeish to L.L. Bolton, 12 December 1925; all in PAC, Geological Survey of Canada, RG 45/33/4. S.C. Ells to H. Stutchberry, 23 May 1925, PP, 174; T. Draper to W.W. Cory, 27 August 1930; T. Draper to C. Camsell, 9 September 1930; both in RG 45/34/6; R.C. Wallace to T. Draper, 26 November 1930,

RCW, 3/2/6/2.

12. K. Clark, "Preliminary Report re: Bituminous Sand Sidewalks Laid by Thos. Draper," November 1930, RCW, 3/2/6/2; L.E. Drummond to K.A. Clark, 27 May 1939; K. Clark to L.E. Drummond, 30 May 1939; both in UAA, MM, 300.

13. S.C. Ells to A.W.G. Wilson, 17 April 1925, RG 45/33/4; T. Draper to C. Camsell, 9 September 1930, RG 45/34/6; K. Clark, Diary, mss., entry for 7 September 1927, KAC, Box 1.

14. K.A. Clark, "Report ..." October 1930, UAA, RCA, 80/1/2/3-6; Ells, "Bituminous Sands of Northern Alberta," (1926), 173-234. Ells's report enables identification of twelve or thirteen experiments (he is unclear about a couple). To these should be added the work of W.P. Hinton, Jacob Absher, Robert Fitzsimmons, and Max Ball and his associate, J.M. McClave.

15. Ells, "Bituminous Sands of Northern Alberta," (1926), 194-95.

16. Ibid., 203-204 and plate XLIII, 205-206, 230-31, 233-34, 206-207; *Alberta Oil Examiner*, 28 March 1926.

17. Ells, "Bituminous Sands of Northern Alberta," (1926), 210. J.B. Lindsay to H.M. Tory, 7 and 27 October 1920; H.M. Tory to J.B. Lindsay, 16 October 1920; both in HMT, 906-5a-2.

18. Ells, "Bituminous Sands of Northern Alberta," (1926), 220, 224.

19. Ibid., 200-201.

20. R. Caulfield to H. Stutchbury, 9 July 1926, PP, 174; K. Clark, "Report ...," October 1930, RCA 80/1/2/3-6; Savary, Fenerty and McLaurin to J.E. Brownlee, 5 August 1926, 19 October 1926, 31 January 1927, PP, 174; *Alberta Oil Examiner*, 12 June 1926, 7 August 1926, 14 August 1926, 28 August 1926, *Western Oil Examiner*, 30 April 1927.

21. K. Clark, "Progress Report," 30 November 1927, RCA, 80/1/2/3-6; K. Clark, Diary, mss., entries for 10, 16 September 1927, 3 October 1929, 11 May 1930, 8, 22 June 1930, KAC, Box 1; K. Clark to R.C. Wallace, 7 February 1930,

RCW, 3/2/6/2; *Western Oil Examiner*, 7 July 1928.

22. W.P. Hinton to H. Stutchbury, 5 July 1929; K. Clark to W.P. Hinton, 12 and 16 July 1929; K. Clark to H. Stutchbury, 16 June 1929; K. Clark to R.C. Wallace, 23 July 1929; all in RCW, 3/2/6/2.

23. C. Camsell to R.C. Wallace, 29 July 1929; R.C. Wallace to W.P. Hinton, 12 September 1929; W.P. Hinton to R.C. Wallace, 20 September 1929; all in RCW, 3/2/6/2; K. Clark, Diary, mss., entry for 29 October 1929, KAC, Box 1; K. Clark, "Progress Report," 6 December 1929; R.C. Wallace to C. Camsell, 9 December 1929; K. Clark to R.C. Wallace, 21 February 1929; all in RCW, 3/2/6/2.

24. Cf. N. Edwards to W.P. Hinton, 6 November 1929; H. Stutchbury to J.E. Brownlee, 22 November 1929; J.E. Brownlee to R.C. Wallace, 25 November 1929; cf. K. Clark, "Progress Report," 6 December 1929; all in RCW, 3/2/6/2.

25. Biographical information on Fitzsimmons's life is voluminous in amount if somewhat sparse on detail in his extraordinary collection of personal and business papers at the Provincial Archives of Alberta.

26. Drilling Reports 1927 to 1930, RCF, 11/299 and 11/310; R.C. Fitzsimmons to H.H. Rowatt, 20 December 1927, RCF, 11/288; Duncan Rule to R.H. MacMicking, 13 December 1927, RCF, 11/286; K. Clark, Diary, mss., entry for 18 September 1927, KAC, Box 1; H.H. Rowatt, memo. 22 June 1928, RCF, 11/316; R.C. Fitzsimmons to A. Norquay, 16 April 1928, RCF, 8/328.

27. Drilling Report for 22 February 1930, RCF, 11/310; K. Clark, Diary mss., entry for 26 June 1930, KAC, Box 1; K. Clark, "Progress Report," October 1930, RCA, 80/1/2/3-6; S.C. Ells quoted in Canada, Mines Branch, "Investigation of Mineral Resources in Canada, 1930," Mines Branch *Report No.* 723 (Ottawa: King's Printer, 1931), 7-10.

28. K. Clark, "Progress Report," October 1931, RCA, 80/1/2/3-6; Max Ball to C.H. Mitchell, 25 March 1935, RCW, 3/2/6/2; R. Fitzsimmons, "What Happened to International Bitumen?," May 1953, typescript, RCF, 16/543.

29. K. Clark, "Progress Report," October 1931, RCA, 80/1/2/ 3-6; R.C. Fitzsimmons to R. MacMicking, 4 April 1931, RCF, 11/286.

30. R.C. Fitzsimmons, "Report on Operations," 18 November 1931, RCF, 11/319.

31. A.W. Haddow to R.C. Fitzsimmons, 29 June 1932, RCF, 16/541.

32. R.C. Fitzsimmons to J. Harvie, 27 August 1932; R.C. Fitzsimmons to C. Stewart, 10 June 1930; both in RCF, 8/328.

33. R.C. Fitzsimmons to F.J. Falconer, 8 March 1933, RCF 11/289; R.C. Fitzsimmons to S.C. Ells, 6 May 1933, RCF 9/245b; S.C. Ells to R.C. Fitzsimmons, 4 and 25 August 1933, RCF, 9/254b; P.G. Davies to R.C. Fitzsimmons, 31 December 1934, RCF, 11/318; R.C. Fitzsimmons to L. Ashenhurst, 22 November, 15 December 1934, RCF, 11/318. On the Bitumount plant, see J.G. Knighton to C.T. Oughtred, 17 August 1938, CM&S, File 165ab; cf. Fitzsimmons, "What Happened to International Bitumen?," where Fitzsimmons claims that his firm produced 250 tons per day in 1930—a slight exaggeration. (Knighton's letter and estimates ended up with Fitzsimmons, too, fn. 43).

34. For example, R.C. Fitzsimmons to S.C. Ells, 11 October 1935, RCF, 9/254b. According to Fitzsimmons, a $150,000 financing offer was available to International Bitumen in 1939. Provincial officials, he stated, refused to verify the firm's valuable properties and process, thus scotching the chance: R.C. Fitzsimmons to N.E. Tanner, 10 January 1939, R.C. Fitzsimmons to R. MacMicking, 16 June 1939, 13 May 1940, all in RCF, 10/272a.

35. R.C. Fitzsimmons to H. Everard, 19 June 1934; H. Everard, 5 June 1934; R.C. Fitzsimmons to H.F. Everard, 10 May and updated 1936; H. Everard to R.C. Fitzsimmons, 5 and 12 May 1936; all in RCF, 10/256a.

36. See note 35 and the following: H. Everard, Report on International Bitumen, 17 July 1936, RCF, 11/290; R.C.

Fitzsimmons, "Memorandum," December 1936, RCF, 16/539.

37. H. Everard to R.C. Fitzsimmons, 20 October 1936; V. Engman to R.C. Fitzsimmons, 20 October 1936; both in RCF, 16/539.

38. H. Everard to R.C. Fitzsimmons, 18 February 1937, RCF, 10/256a; H. Everard to R.C. Fitzsimmons, 11, 18, 24 May 1937, RCF, 10/256c; H. Everard to R.C. Fitzsimmons, 25 June 1937, RCF, 10/254a; L. Romanet to Max Ball, 20 June 1937, LAR, 7/1/4.

39. H. Everard to R.C. Fitzsimmons, 1 and 7 July 1937, 7 August 1937, 3 September 1937, 8 September 1937 (collect telegram), RCF, 10/256c. On the labour disputes at Bitumount see Louis Romanet to Max Ball, 6 August 1937, RCF, 9/238a. W.E. Adkins reported on plant conditions to the Alberta Department of Lands and Mines in 1937, RCF, 9/238a.

40. H. Everard to R.C. Fitzsimmons, 29 October and 22 November 1937, RCF, 10/256a; H. Everard to R.C. Fitzsimmons, 8 and 23 January 1938, 3 April 1938; RCF, 10/256c; R.C. Fitzsimmons to H. Everard, 23 January 1938, RCF, 10/256c. A. Carsten to R. Fitzsimmons, 18 March 1939; R. Fitzsimmons to A. Carsten, 20 April 1939, both in RCF, 10/256d; R. Fitzsimmons to R. MacMicking, 25 November 1939, RCF, 10/272a; Parlee, Smith, Clement and Parlee to M. Falconer, 2 July 1943, RCF, 18/676.

41. R. Fitzsimmons to Frank Badura, 26 January 1938, RCF, 11/286; "Expenses of International Bitumen," 1 January, 30 June 1943, RCF, 15/479; Parlee, Smith, Clement and Parlee to M. Falconer; 2 July 1943, RCF, 18/676; c.f. V. Engman, who claimed wages: R. Fitzsimmons to R. MacMicking, 22 January 1938, RCF, 11/286.

42. In general see RCF 9/238c, but especially the following: W.E. Adkins to R.C. Fitzsimmons, 2 October 1937, and 2 June 1938, RCF, 9/238.

43. "Mineral Developments in Northern Alberta," Alberta Department of Lands and Mines, 27 May 1938, PP, 792.

W.E. Adkins to R. Fitzsimmons, 8 and 10 June 1938, 2 July 1938; J.G. Knighton to C.T. Oughtred, 17 August 1938, both in RCF, 9/238; R. MacMicking to R. Fitzsimmons, 26 November 1938, RCF, 10/272a.

44. R.C. Fitzsimmons to R. MacMicking, 22 January 1938, RCF, 11/286; W.E. Adkins to R. MacMicking, 6 September, 29 October, 11 November 1938; R. MacMicking to W.E. Adkins, 16 November 1938, 17 March 1939, 13 April 1939, 20 August 1940; both in RCF, 9/238.

45. R. Fitzsimmons to W.E. Adkins, 16 March and 16 May 1941; W.E. Adkins to R. Fitzsimmons, 18 May and 3 December 1941; both in RCF, 9/238.

46. Employees' Time Sheets for 1938, in RCF, 15/434.

47. R. MacMicking to R. Fitzsimmons, 26 November 1938, RCF, 10/272a.

48. R. MacMicking to R. Fitzsimmons, 14 December 1938, RCF, 15/480; "International Bitumen Co.," statement re: R. MacMicking, RCF, 15/482; R. Fitzsimmons to R. MacMicking, 17 December 1938, 3 June 1940; R. MacMicking to R. Fitzsimmons, 23 January 1940, 28 May 1940, both in RCF, 10/272a.

49. G.J. Knighton to C.T. Oughtred, 17 August 1938, RCF, 10/257 and CM&S, File 165ab.

50. J.W. Fairlie, "Report on International Bitumen," 22 September 1938; R.C. Fitzsimmons to J.W. Fairlie, 18 January 1939; both in RCF, 10/257; S.C. Ells to B.F. Haanel, 31 December 1939, PAC, Canada, Mineral Resources Branch, RG 87/111/540; S.C. Ells to R. Fitzsimmons, 5 December 1933, 27 March and 9 May 1934, RCF, 9/254b.

51. L.C. Stevens, "Report on Oil Sands Ltd. Property," 27 March 1943, RCF, 16/507.

52. S.C. Ells, Memorandum: International Bitumen, 26 April and 14 August 1943, PAC, Canada, Department of Energy, Mines and Resources, RG 21/40/3; R. Fitzsimmons to W.E. Adkins, 23 June 1943, RCR, 9/247a; L. Champion to R. Fitzsimmons, 22 June 1944, RCF, 9/247a; R. Fitzsimmons to L. Champion, 24

February, 3 July, 2 August, 7 September, 4 October 1944, RCF, 9/247a; R. Fitzsimmons to Guy Hamond, 15 July 1944, RCF, 10/263a. International Bitumen's dormancy is made clear in both Fitzsimmons's Papers, file 9/247a and in the federal government's report on Mining for 1944: Canada, Bureau of Mines, "The Canadian Mineral Industry in 1944," Bureau of Mines *Report No.* 815, (Ottawa: King's Printer, 1945), 50-51.

53. Ells, "Bituminous Sands of Northern Alberta," (1926), 25-26; Max W. Ball, "Development of the Athabaska Oil Sands," *Transactions*, Canadian Institute of Mining and Metallurgy 44 (1941), 76. See also the Energy Resources Conservation Board file on W.P. Hinton, which notes the 1921 work McClave did analysing the oil sands samples which Hinton was also interested in: PAA, ERCB 6/138.

54. Wallace Pratt, "Max W. Ball," *Bulletin*, American Association of Petroleum Geologists 39, no. 5 (May 1955): 975-80.

55. K. Clark, "Memo. Re: Max Ball and Jones of Denver...," 4 March 1930; Charles Stewart to R.C. Wallace, 31 May and 4 June 1930; both in RCW, 3/2/6/2.

56. Max Ball to R.C. Wallace, 14 June 1939; R.C. Wallace to Max Ball, 19 June 1930; both in RCA 80/1/2/4-1; K. Clark. "Memo....," 4 March 1930, RCW, 3/2/6/2; Max Ball to George Hoadley, 8 August 1930; Max Ball to Research Council of Alberta, 30 August 1930; both in RCW, 3/2/6/2.

57. K. Clark, "Report...," October 1930, RCA, 80/12/2/3-6; Max Ball to J.E .Brownlee, 8 June 1931, RCA, 80/1/2/4-1.

58. Max Ball to J.E. Brownlee, 8 June 1931, RCA, 80/1/2/4-1; K. Clark to R.C. Wallace, 8 April 1931, RCW, 3/2/6/2; Max Ball to R.G. Reid, 9 August 1933, RCA, 80/1/2/4-1; Mines Branch, Memorandum, 28 December 1934; John McLeish, Memorandum, 26 December 1933; both in RG 45/34/7; Max Ball to Minister of Interior (T.A. Crerar), 24 August 1937, RG 21/43/1; see also Ball's "Development of the Athabaska Oil Sands," 76-77.

59. S.G. Blaylock to Professor N. Pitcher, 10 January 1931; R.C. Wallace to S.G. Blaylock, 13 January 1931; both in RCA, 80/1/2/4-1; K. Clark to S.G. Blaylock, 16 January 1931, RCW, 3/2/6/2. The work done by Consolidated Mining,

much of it under Karl Clark's supervision or else by his former students, is summarized in K.A. Clark, "The Hot Water Method for Recovering Bitumen" typescript, Sullivan Concentrator, Chapman Camp, British Columbia 1 September 1939. In *Papers relating to Alberta Bituminous Sand*, 1939-50, by Karl A. Clark. Bound in the University of Alberta Library.

60. W.P. Campbell, "Report on Abasand Oils Plant," 9 November 1936, RG 21/43/1; Max Ball, in an unpublished essay likely often presented as a speech, provided similar evidence of his plant's activities: "Oil in the Oil Sands of Northern Alberta," typescript October 1936, LAR, 7/1/4.

61. Max Ball to Minister of Interior, 5 September 1936; Max Ball to Minister of Interior, 24 August 1937; Max Ball to H.S. Dunn, 1 September 1937; all in RG 21/43/1. See also: Alberta, Department of Lands and Mines, "Mineral Developments in Northern Alberta," 27 May 1938, typescript, PP, 792.

62. Max Ball to T.A. Crerar, 15 June, 22 July 1939; Max Ball to J. MacKinnon, 15 June 1939; T.A. Crerar to Max Ball, 5 July 1939; all in RG, 21/43/1; Max Ball to L.E. Drummond, 7 December 1938; C.T. Oughtred to F.H. Chapman, 7 December 1938; Max Ball to F.H. Chapman, 7 December 1938; all in PAA, CM&S, 1.

63. See Canada, Bureau of Mines, "The Canadian Mineral Industry in 1944," 50-51; K. Clark to L.A. Westman, 16 February 1940, MM, 313; Max Ball to C. Camsell, 19 December 1939; Max Ball to R. Gibson, 5 October 1940; J. McLeish to R. Gibson, 17 October 1940; all in RG, 21/43/1; Max Ball to R.W. Diamond, 2 August 1940; C.T. Oughtred to R.W. Diamond, 13 July 1940; both in CM&S, 2.

64. G.R. Cottrelle to C.D. Howe, 4 February 1942, RG, 21/40/2.

65. See Ball, "Development of the Athabaska Oil Sands," 91; Earle Gray, *Wildcatters: The Story of Pacific Petroleums and West Coast Transmission* (Toronto: McClelland and Stewart, 1982), 45-53, 73.

66. H.G.J. Aitken, *American Capital and Canadian Resources* (Cambridge, Mass.: Harvard University Press, 1961).

67. On the industry's condition see: Gray, *Wildcatters*; David H. Breen, ed. *William Stewart Herron: Father of the Petroleum Industry in Alberta* (Calgary: Alberta Records Publication Board, 1984), xxxvii ff.; August W. Giebelhaus, *Business and Government in the Oil Industry: A Case Study of Sun Oil, 1876-1945* (Greenwich, Conn.: JAI Press, 1980), 198-233.

IV: ABASAND AS A WARTIME PROJECT

1. The two best general accounts of Canada's economic war effort set into the context of the period are J.L. Granatstein, *Canada's War: The Politics of the Mackenzie King Government, 1939-45* (Toronto: Oxford University Press, 1975) and Robert Bothwell and William Kilbourn, C.D. *Howe: A Biography* (Toronto: McClelland and Stewart, 1979). See also: J. de N. Kennedy, *History of the Department of Munitions and Supply*, 1 (Ottawa: E. Cloutier, 1950).

2. Canada, House of Commons, "Special Committee on Reconstruction and Re-establishment," Minutes and Proceedings, no. 33, 30 November 1943, PAC, Canada, Department of Energy, Mines and Resources, RG 21/40/3; Oil control, untitled draft memorandum on Canadian consumption of oil and supply, 1940-42, PAC, Canada, Department of Munitions and Supply, RG 28-A/250/196/10/6. Max Ball, *This Fascinating Oil Business* (Indianapolis and New York: The Bobbs Merrill Company, 1940).

3. "Bituminous Sand Permit No. 1," agreement with Max Ball, 23 May 1930; "Bituminous Sand Permit No. 2," agreement with W.P. Hinton, 10 June, 1930; both in PAC, Geological Survey of Canada, RG 45/34/6; A.W.G. Wilson, "Memorandum, #2, Re: Alberta Claim to Federal Reserve in Bituminous Sands," 12 December 1934, RG 45/34/7; Charles Camsell to R.A. Gibson, 20 May 1930, RG 45/34/6; L.L. Bolton to R.A. Gibson, 13 May 1930, RG 45/34/5. Memorandum "Estimate: Dominion Government disbursements related to Bituminous Sands of Northern Alberta, 1913-22," RG 21/40/3; the total amount spent 1913-22 was $105,480.12, of which $29,954.54 was for capital expenses.

4. See Chapter III for a fuller description of this. Max Ball to

L.E. Drummond, 7 December 1938; C.T. Oughtred to F.H. Chapman, 29 July 1938; both in PAA, CM&S, I; G.J. Knighton to C.T. Oughtred, 17 August 1938, CM&S, 65ab.

5. For example, Karl Clark to Dr. Gustave Egloff, 11 April 1938, CM&S, 8; K.A. Clark to G.J. Knighton, 12 April 1938, CM&S, 7; See also K.A. Clark to W.G.Jewett, 26 May 1939, K.A. Clark to S.G. Blaylock, 6 October 1939, both in UAA, MM, 297.

6. W.G. Jewett to S.G. Blaylock, 19 January 1939; S.G. Blaylock to W.E. Stavert, 16 January 1939, both in CM&S, 9a.

7. See Richard Diubaldo, "The Canol Project in Canadian-American Relations," Canadian Historical Association, *Historical Papers* (1977), 179-95.

8. Ibid., 183-5; Granatstein, *Canada's War*, 321-23; cf. C.D. Howe's view as noted in note 32 below.

9. G.R. Cottrelle to C.D. Howe, "Report on Bituminous Sand Deposit," 4 February 1942, RG 21/40/2.

10. John McLeish to R.A. Gibson, 17 October 1940, RG 21/43/1; A.L.Johannson to Max Ball, 30 August 1940, CM&S, 3; W.G. Jewett to R.W. Diamond, 13 July 1940, CM&S, 2; Memorandum, CM&S Legal Department to Mines Department, 30 September 1942, CM&S, 3.

11. G.A. Young to F.C.C. Lynch, 24 January 1942; G.A. Hume, "Report"; E.S. Martindale to W.B. Timm, 4 March 1942; all in RG 21/40/2.

12. Max Ball to Nesbitt, Thomson Co., 28 February 1942; P.A. Thomson to T.A. Crerar, 24 February 1942, both in RG 21/40/2. See also: W.B. Timm to A.J. Nesbitt, 20 April 1943; H.S. Dunn to W.B. Timm, 19 April 1943; W.B. Timm to H.S. Dunn, 21 April 1943; all in RG 21/43/1.

13. W.B. Timm to Charles Camsell, 23 April 1942, and Report on "Reproduction and Refining of 10,000-bbl./day of Bitumen...," RG 21/40/2.

14. S.G. Blaylock to C.D. Howe, 27 March 1942; J.R. Donald to W.S. Kirkpatrick, 5 May 1942; both in CM&S, 9a.

15. Report, "Committee to Investigate the Oil Sands of the

Athabaska River," CM&S, 95b and in PAC, Canada, Mineral Resources Branch, RG 87/88/659, and also correspondence in CM&S, 33 (1942).

16. G. Cottrelle to C.D. Howe, 2 June 1942, PAC, C.D. Howe Papers, 34/35. "Minutes of Meeting to Discuss Kirkpatrick Report," 11 June 1942, C.D. Howe Papers, 34/3; C.D. Howe to S.G. Blaylock, 10 June 1942, CM&S, 50; C.D. Howe to G. Cottrelle, 13 June 1942; C.D. Howe to S.G. Blaylock, 25 June 1942; C.D. Howe to T.A. Crerar, 17 July 1942; all in C.D. Howe Papers, 34/35. See also: Oil Control, "Report on World Oil Situation," RG 28-A/250/196/16, 1942.

17. W.S. Kirkpatrick to R.W. Diamond, 9 July 1942, CM&S, 11a; W.S. Kirkpatrick to A.D. Turnbull, 22 and 31 July 1942, CM&S, 9b.

18. A.D. Turnbull to W.S. Kirkpatrick, 9 and 22 July 1942, CM&S, 9b; Turnbull to Kirkpatrick, 22 August 1942, quoted in CM&S, 6a; Turnbull to Kirkpatrick, 18 August 1942, CM&S, 9b.

19. A.D. Turnbull to W.S. Kirkpatrick, 22 August CM&S, 6a; Turnbull to Kirkpatrick, 4 September 1942, CM&S 9b; C.D. Howe to S.G. Blaylock, 5 September 1942, CM&S, 12a.

20. W.S. Kirkpatrick to S.G. Blaylock, 15 September, CM&S, 12a; W.S. Kirkpatrick to A.D. Turnbull, 11 September 1942, CM&S, 9b.

21. A.D. Turnbull to W.S. Kirkpatrick, 7 and 19 October 1942; P.L. Barron to G.A. Wallinger, 4 November 1942; both in CM&S, 11a/b. A.D. Turnbull to W.S. Kirkpatrick, 21-22 October 1942, CM&S, 9b.

22. W.S. Kirkpatrick to S.G. Blaylock, 6 November 1942, CM&S, 12a.

23. A.D. Turnbull to W.S. Kirkpatrick, 16-17 November 1942, CM&S, 9b.

24. W.S. Kirkpatrick to S.G. Blaylock, 1 December 1942, CM&S, 31; W.S. Kirkpatrick, "Memorandum of Conversations," 1-3 December 1942, CM&S, 26; W.S. Kirkpatrick to W.G. Jewett, 4 December 1942, CM&S, 31.

25. A.J. Nesbitt to C.D. Howe, 20 November 1942; A.J. Nesbitt

to W.B. Timm, 16 October 1942; Max Ball to C.D. Howe, 21 October 1942; A.J. Nesbitt to G.R. Cottrelle, 13 and 18 November 1942; all in C.D. Howe Papers, 34/35.

26. G.J. Knighton, "Athabaska Oil Sands Investigation...," CM&S, 118.

27. Ibid. See also: W.S. Kirkpatrick to W.B. Timm, 24 February 1943; W.B. Timm to W.S. Kirkpatrick, 1 March 1943; both in CM&S, 6a; W.S. Kirkpatrick to S.G. Blaylock, 31 March 1943, CM&S, 9c.

28. W.S. Kirkpatrick to W.B. Timm, 16 November 1942, C.D. Howe Papers, 34/35. See also Kirkpatrick's Summary of their oil sands involvement: Kirkpatrick to R.W. Diamond, 20 May 1943, CM&S, 32.

29. A.J. Nesbitt to C.D. Howe, 18 November 1942; W.B. Timm to A.J. Nesbitt, 20 April 1943; A.J. Nesbitt to W.B. Timm, 27 April 1943; Memorandum, Accounts Payable, 31 March 1943; all in RG 21/43/1. W.S. Kirkpatrick to R.E. Stavert, 17 June 1944, CM&S, 4. A.J. Nesbitt to Max Ball, 8 January 1943, PAC, Earl Smith Papers; G.R. Cottrelle to Earl Smith, 22 February 1943, Earl Smith Papers; Earl Smith, Laboratory Report, 3 December 1926, Earl Smith Papers; Karl Clark to M.W. Lyons, 29 September 1942, MM, 353.

30. Earl Smith to G.R. Cottrelle, 23 February 1943, RG 21/43/1; Earl Smith, "Estimate," 2 March 1943, RG 87/88/634; R.A. Hanwright to Earl Smith, 1 March 1943, Earl Smith Papers.

31. G. Cottrelle to Earl Smith, 20 January, 20 February 1943, Earl Smith Papers.

32. C.D. Howe to S.G. Blaylock, 5 September 1942; G. Cottrelle to C.D. Howe, 16 December 1942; C.D. Howe to G. Cottrelle, 18 December 1942; C. Camsell to C.D. Howe, 3 December 1942; G. Cottrelle to C.D. Howe, 19 January 1943; all in C.D. Howe Papers, 34/35. On Canol see Diubaldo, "The Canol Project."

33. C.D. Howe to G. Cottrelle, 22 January 1943, C.D. Howe Papers, 34/35.

34. de N. Kennedy, History of the Department of Munitions and

Supply, 1, 519-20; *Western Oil Examiner*, 17 April, 1 May, 15 May 1943; Philip Smith, *The Treasure Seekers: The Story of Home Oil* (Toronto: Macmillan of Canada, 1978), 96-100.

35. See Privy Council 3058, 15 April 1943, RG 21/43/1; W.H. Norrish, Re: Abasand Oils Ltd. Agreement, 26 April 1943; Max Ball, "Letter to Shareholders," 1 April 1943; both in RG 21/43/1. See also *Hansard*, 14 April 1943, for C.D. Howe's statement.

36. P.V. Rosewarne to G.R. Cottrelle, 12 February 1943, RG 21/43/1.

37. Earl Smith to W.S. Kirkpatrick, 1 April 1943, CM&S, 126; S.G. Blaylock to W.S. Kirkpatrick, 10 April 1943; W.S. Kirkpatrick to W.G. Jewett, A.D. Turnbull and L. Telfer, 22 April 1943, both in CM&S, 9c.

38. B.F. Haanel to W.B. Timm, 17 April 1943, RG 21/43/1; P.V. Rosewarne to W.B. Timm, 27 April 1943, RG 21/44/2; W.H. Norrish, Memorandum, 4 May 1943, RG 21/44/2.

39. S.C. Ells to W.B. Timm, 16 August 1943; P.V. Rosewarne to G. Cottrelle, 31 August 1943; both in RG 21/44/3.

40. G. Cottrelle to P.V. Rosewarne, 11 September 1943; G.B. Webster to G. Cottrelle, 16 September 1943; both in RG 21/44/3.

41. G. Cottrelle to John Irwin, 17 September 1943; H.S. Dunn to W.B. Timm, 30 September 1943; both in RG 21/44/3; Earl Smith to W.S. Kirkpatrick, 27 September 1943, CM&S, 9c.

42. H.S. Dunn to W.B. Timm, 30 September 1943, RG 21/44/3; P.V. Rosewarne to G. Cottrelle, 4 October 1943, RG 21/44/4.

43. W.B. Timm to G.B. Webster, 7 October 1943; G.B. Webster to W.B. Timm, 14 October 1943; H.L. Tamplin to G.B. Webster, 16 October 1943; W.B. Timm to G.R. Cottrelle, 5 November 1943; all in RG 21/44/4.

44. H.L. Tamplin, "Progress Report #6, Abasand Oils," 5 March 1944; Minutes, Abasand Conference, Ottawa, 30 March 1944; both in RG 21/44/6.

45. H.L. Tamplin, "Progress Report #7," 5 April 1944; G.B. Webster to W.B. Timm, 18 April 1944; both in RG 21/44/6.

46. W.B. Timm to John Irwin, 29 July 1944, RG 21/44/7; G.B. Webster to W.B. Timm, 15 August 1944, RG 21/44/8.

47. Arthur MacNamara to C.W. Jackson, 22 August 1944 and 20 September 1944; A. MacLachlan to Arthur MacNamara, 18 September 1944; W.H. Norrish to W.B. Timm, 21 August 1944; all in RG 21/4445/8. See also: H. Tamplin's Reports quoted in Note 49 below.

48. W.B. Timm to C. Camsell, 20 September 1944; G.S. Hume to W.B. Timm, 28 September 1944; G. Cottrelle to W.B. Timm, 3 October 1944, all in RG 21/45/8.

49. H. Tamplin, "Progress Reports," #13, 6 October 1944; #12, 5 September 1944; #11, 4 August 1944; RG 21/45/8; #14, 4 November 1944, RG 21/45/9.

50. C.M. Sprague to W.B. Timm, 23 October 1944, RG 21/45/8; G.B. Webster to W.B. Timm, 15 December 1944; H.A. Brown to B.J. Block, 3 January 1945; both in RG 21/45/9.

51. H. Tamplin, "Progress Report #16," 5 January 1945; G.B. Webster to W.B. Timm, 14 January 1945; W.B. Timm to John Irwin, 15 February 1945; all in RG 21/45/9.

52. H. Tamplin, "Progress Report #16," 5 January 1945, RG 21/45/9. See also: G. Cottrelle to W.B. Timm, 13 October 1944, RG 21/45/8.

53. H. Tamplin, "Progress Report #18," 10 March 1945, RG 21/45/10.

54. P.D. Hamilton to President and Directors, Abasand Oils, 11 April 1945; W.B. Timm to C.W. Jackson, 12 April 1945; both in RG 21/45/10.

55. Compiled from P.D. Hamilton, "Progress Report #19, #20, #21," 11 April, 17 May, 11 June 1945, RG 21/45/10 and 11. See also: D.A. Duff to Petroleum and Natural Gas Conservation Board of Alberta, 11 May 1945, RG 21/45/10.

56. Compiled from P.D. Hamilton, "Progress Report #19, #20, #21, 11 April, 17 May, 11 June 1945, RG 21/45/10 and 11.

57. P.D. Hamilton to G.B. Webster, 20 June 1945, RG 21/45/11.

58. G.B. Webster to W.B. Timm, 29 June 1945, RG 21/45/11.

59. *Globe and Mail,* 16 June 1945; *Financial Post,* 21 July 1945; W.H. Fallow, statement in *Winnipeg Free Press,* 12 November 1943; Statement of T.A. Crerar, 16 November 1943 (Refuting a Fallow Statement) RG 21/40/3; W.D. Fallow, statement in *Edmonton Journal,* 14 March 1944.

60. P.D. Hamilton, Report, 3 August 1945; P.D. Hamilton to G.B. Webster, 29 August 1945; both in RG 21/45/12.

61. C.D. Howe to J.A. Glen, 22 August 1945, RG 21/45/11.

62. J.A. Glen to C.D.Howe, 28 August 1945; W.B. Timm to John Irwin, 25 August 1945; both in RG 21/45/11.

63. Report for Abasand Oils Ltd., by John Irwin and G.B. Webster, 7 September 1945, RG 21/45/12; W.H. Norrish, Memorandum to Deputy Minister, 12 October 1945, RG 87/88/635; W.B. Timm to C.W. Jackson, 12 October 1945, RG 21/45/12.

64. G.B. Webster to W.B. Timm, 17 September 1945; Report on Operations, Abasand Oils, 10 October 1945; Abasand Oils to W.B. Timm, 12 October 1945; all in RG 21/45/12.

65. G.B. Webster, Report, 10 October 1945, RG 21/45/12.

66. G.B. Webster to W.B. Timm, 11 October 1945; J.A. Glen to John Irwin, 19 October 1945; W.B. Timm to G.B. Webster, 20 October 1945; C.D. Howe to J.A. Glen, 28 November 1945; John Irwin to J.A. Glen, 30 November 1945; C.D. Howe to J.A. Glen, 5 June 1946; C.D. Howe to J.A. Glen, 5 July 1946; J.A. Glen to John Irwin, 24 July 1946; all in RG 21/46/14.

67. See J.N. McDougall, *Fuels and the National Policy* (Toronto: Butterworths, 1982), Chapter 4. David H. Breen has begun this useful study in David H. Breen, "Anglo-American

Rivalry and the Evolution of Canadian Petroleum Policy to 1930," *Canadian Historical Review* 62 (Summer 1981): 283-303.

68. Memorandum, 1 November 1946, RG 21/46/18; H.S. Dunn to W.B. Timm, 8 October 1946, RG 21/46/14; Cabinet Memorandum, 16 July 1948, RG 21/46/16. W.H. Norrish, memorandum, 23 March 1954, RG 21/46/18.

69. The summary of Canada's sterling war effort is to be found in Granatstein's *Canada's War,* 419-24. Cf. Darlene J. Comfort, *The Abasand Fiasco: The Rise and Fall of a Brave Oil Sands Extraction Plant* (Edmonton: Friesen Printers, 1980); and J.J. Fitzgerald, *Black Gold with Grit; the Alberta Oil Sands* (Sidney, B.C.: Gray's Publishing Ltd., 1978).

V: BITUMOUNT

1. Cf. Larry Pratt, *The Tar Sands: Syncrude and the Politics of Oil* (Edmonton: Hurtig Publishers, 1976), 27-29, 30-32, 37-48. He suggests that oil sands development was hindered by government policy which rejected the possibility of public ownership and yet impeded private development by reversing the oil sands for large-scale private enterprise. Pratt simultaneously demands a policy of government pressure for quick resource development and criticizes reliance on conventional reserves for development. Pratt disdains the pricing of oil and markets as factors.

2. William Aberhart to C.D. Howe, 3 July 1942; E.C. Manning to C.D. Howe, 7 January 1944, both in PP, 811B, W.A. Fallow to J.A. Glen, 12 December 1945, PAC, Canada, Department of Energy, Mines and Resources, RG 21/45/13.

3. See J.R. Mallory, *Social Credit and the Federal Power in Canada* (Toronto: University of Toronto Press, 1965), 155-63; C.B. Macpherson, *Democracy in Alberta: Social Credit and the Party System* (Toronto: University of Toronto Press, 1953), 200-212.

4. R.C. Fitzsimmons to W.E. Adkins, 23 June 1943, PAA, RCF, 9/238; R.C. Fitzsimmons to L.R. Champion, 14 May 1944;

L.R. Champion to R.C. Fitzsimmons, 4 April, 22 June 1944; both in RCF, 9/247a.

5. K.A. Clark to N.E. Tanner, 27 September 1944, PP, 813.

6. Lloyd Champion to Government of Alberta, 13 October 1944, PP, 813.

7. E.C. Manning to Oil Sands Ltd., 8 December 1944, PP, 813; Order in Council, 1885/44, 6 December 1944, PP, 1601.

8. Karl Clark to Merle Bancroft, 3 September 1945, UAA, MM, 411.

9. Karl Clark to Professor E.O. Lilge, 23 July 1946, MM, 411.

10. Karl Clark, "Progress Report, Bituminous Sands Investigations," May 1946, PP, 1505B; Minutes, Annual General Meeting of the Research Council of Alberta, 21 November 1946, PP, 1505B.

11. S.M. Blair to K.A. Clark, 24 June 1946, MM, 411.

12. K.A. Clark to S.M. Blair, 20 July, MM, 411.

13. K.A. Clark to S.M. Blair, 20 July, MM, 411.

14. Karl Clark to S.M. Blair, 9 August 1947, MM, 444.

15. Karl Clark to S.M. Blair, 9 August 1947, MM, 444.

16. L.R. Champion to E.C. Manning, 21 October 1947, PP, 1348.

17. Order in Council, 199/49, 15 February 1949; K. McKenzie to Peter Elliot, 25 January 1949; both in PAA, PWM, 1, 1949.

18. George Clash to D.B. Macmillan, 1 June 1948, PP, 1645; Karl Clark to A. Wolverton, 27 May 1948, MM, 460.

19. Staff, Oil Sands Project, "Report to the Board of Trustees on the Oil Sands Project From Inception to December 31, 1948," Alberta, *Sessional Paper* No. 53 (1949), 1-2.

20. Staff, Oil Sands Project, "Report to the Board of Trustees, 1948," 5-6, 17-19, 28-30.

21. Staff, Oil Sands Project, "Report to the Board of Trustees, 1948," 11-12, 21-23, 26-27.

22. See D.B. Macmillan's announcement about the success of Bitumount: *Edmonton Bulletin*, 21 October 1948 and *Calgary Herald*, 22 October 1948, clippings, in PAA, APS, 54.

23. Oil Sands file, PWM, 1949.

24. Board of Trustees, Minutes, 19 February 1949; J.J. McCann to D.B. Macmillan, 21 March 1949; L. Maynard to J.J. McCann, 24 February 1949; all in PWM, 1.

25. W.E. Adkins to D.B. Macmillan, 18 March 1949, PWM, 1949, 1; see Chapter IV on Abasand's operations.

26. W.E. Adkins to D.B. Macmillan, 18 May, 1 June 1949, PWM, 1949, 1.

27. W.E. Adkins to D.B. Macmillan, 16 June, 21 June 1949, PWM, 1949, 1.

28. W.E. Adkins to D.B. Macmillan, 28 and 29 June 1949; W.E. Adkins to Sidney Born, 28 June 1949; both in PWM, 1949, 1.

29. W.E. Adkins to D.B. Macmillan, 29 June 1949, 9 July 1949, PWM 1949, 1.

30. W.E. Adkins to N.E. Tanner, 5 September 1949; W.E. Adkins to D.B. Macmillan, 13 September 1949; both in PWM 1949, 1. On the post-war economy see Robert Bothwell, John English, and Ian Drummond, *Canada Since 1945: Power and Politics and Provincialism* (Toronto: University of Toronto, 1981), 41-42.

31. George Clash to D.B. Macmillan, 5 August 1949; W.E. Adkins to D.B. Macmillan, telegrams, 29 July, 30 July, 10 August 1949; both in PWM, 1. See also the report on the 1949 season's operations, Staff, Oil Sands project, "Report to the Board of Trustees on the Alberta Government Oil Sands Project from January 1, 1949 to December 31, 1949," Alberta *Sessional Paper*, No. 55 (1950), 32.

32. W.E. Adkins to N.E. Tanner, 5 September 1949; W.E. Adkins to D.B. Macmillan, 13 September 1949; both in PWM, 1949, 1.

33. W.E. Adkins to D.B. Macmillan, 17 September 1949, in

PWM, 1949, 1; Karl Clark to W. Fotheringham, 19 September 1949; Karl Clark to Don Picket, 29 August 1950; both in MM, 476 and 499.

34. J.L. Robinson to D.B. Macmillan, 2 September 1949; C.K. Huckvale to D.B. Macmillan, 29 November 1949; both in PWM, 1949, 1.

35. "Summary of Plant Operations, 1949," PWM, 1949, 1; and Staff, Oil Sands Project, "Report to the Board of Trustees, 1949," 4.

36. "Summary of Plant Operations 1949," PWM, 1949, 1; Staff, Oil Sands project, "Report to the Board of Trustees, 1949." See also Karl Clark, "General Information Re: Bitumount Plant, 1949," in UAA, KAC; Research Council of Alberta, "Technical Advisory Committee Report," 11 October 1949, PP, 1894A.

37. On costs see George Clash to D.B. Macmillan, 13 December 1949, PWM, 1949, 1; J.M. Tweedie, statement 29 March 1950, PW, Provincial Marketing Board, 4; Capital and Operational Costs at Bitumount, statement dated 31 March 1955, PP, 1646.

38. Board of Trustees, Minutes, 5 October , 30 October 1949, PWM, 1949, 1.

39. W.E. Adkins to Board of Trustees, Oil Sands Project, 4 October 1949, PWM, 1949, 1, and PP, 1645.

40. Board of Trustees, Minutes, 26 November 1949, PWM, 1949, 1.

41. S.M. Blair to D.B. Macmillan, 30 November 1949; S.M. Blair to J.L. Robinson, 6 December 1949; both in PWM, 1949, 1.

42. W.E. Adkins to D.B. Macmillan, 17 December 1949; G.S. Hume to D.B. Macmillan, 11 October 1949; D.B. Macmillan to G.S. Hume, 2 November 1949; all in PWM, 1949, 1.

43. J.L. Robinson to Ed Nelson, 7 October 1949, PWM, 1949, 1.

44. Eric Hanson, *Dynamic Decade: The Evolution and Effects of the Oil Industry in Alberta* (Toronto: McClelland and Stewart, 1958). Hanson explains clearly the oil strikes of 1947 and 1948. See also Tables 5.2 and 5.3 below for information on

the explosion of production from conventional fields after 1947.

45. Board of Trustees, Minutes, 23 May 1951, PAA, TDM, 1881.

46. S.M. Blair, "Report on the Alberta Bituminous Sands," Alberta *Sessional Paper* No. 49 (1951), Conclusions, 6 *et passim*.

47. Ibid., 6, 7-8, 15 ff.

48. Ibid., 27-28.

49. Ibid., 35.

50. S.M. Blair to J.L. Robinson, 4 and 6 January 1957, PP, 1645.

51. K.A. Clark, ed., *Proceedings*; Athabasca Oil Sands Conference September 1951 (Edmonton: Board of Trustees, Oil Sands Project 1951).

52. N.E. Tanner, "Government Policy Regarding Oil Sands Leases and Royalties," in *Proceedings*, Athabasca Oil Sands Conference, 174-76. *Petroleum Times*, 2 November 1951, 30-32. Cf. Larry Pratt, *The Tar Sands*, 30-32, 37-38.

53. Alberta, Petroleum and Natural Gas Conservation Board, "Alberta Oil Industry," (N.p. selected years following 1948).

54. R.A. McMullen to A.J. Hooke, 8 July, 6 September, 30 September 1948, 12 November 1948, APS, 402; Board of Trustees, Minutes, 23 May 1951, TDM, 1881. On the general attempt at Canadian British economic ties during the post-war period see Bothwell, English and Drummond, *Canada since* 1945, 78-90 and J.L. Granatstein. *A Man of Influence: Norman A. Robertson and Canadian Statecraft 1929-68* (Ottawa: Deneau Publishers, 1981), *passim*.

55. R.A. McMullen to Basil Jackson, 1 March 1951; J. Banfield to R.R. Moore, 6 March 1951; R.A. McMullen to S.M. Blair, 18 May 1951; R.A. McMullen to A.J. Hooke, 23 May 1951; R.A. McMullen to R.P. Bower, 11 June 1951; D.G. Smith to R.A. McMullen, 25 July 1951; R.A. McMullen to D.G. Smith, 20 July 1951; all in APS, Agent-General's Records, 54 and 75. See also D.A. Howes, "Alberta's Bituminous Sands," typescript in TD, General Administration Records,

227/267, viii, 129.

56. Howes, "Alberta's Bituminous Sands," TD, General Administration Records 227/267, i-vii, and 83ff, especially 89-104.

57. Howes, "Alberta's Bituminous Sands," TD, General Administration Records 227/267, 103-104. On the 1950s oil supply and production/price system see Bothwell, English and Drummond *Canada since* 1945, 418-23.

58. Arthur M. Johnson, *The Challenge of Change: The Sun Oil Company, 1945-1977* (Columbus: Ohio State University Press, 1983). See also *Public Accounts of Alberta, 1949-50,* Statement No. 4; Albert L. Danielson, *The Evolution of* OPEC (New York: Harcourt Brace Jovanovich, 1982), 132-35.

59. See Richard Vietor, "The Synthetic Liquid Fuels Programme," *Business History Review* 54, no. 1 (Spring 1980): 1-34.

VI: THE RIDDLE OF THE TAR SANDS

1. Canadian Petroleum Association, *Statistical Handbook* (Calgary: Canadian Petroleum Association, 1980), Tables II-4, VI-I. Geological Survey of Canada, "Oil and Natural Gas Resources of Canada 1983," Geological Survey of Canada *Paper* 83-31, Table 1.

2. James E. Gander and Fred W. Belair, "Energy Futures for Canada," Canada, Energy, Mines and Resources, *Report* EP 78-1 (Ottawa: Supply and Services, 1978).

3. G.A. Stewart, Geological Controls on Distribution of Athabasca Oil Sands Reserves, Research Council of Alberta *Information Report No.* 45 (1963), 15-25.

4. Karl Clark to George S. Hume, 21 August 1952, 30 October 1952, PAC, Canada, Department of Energy, Mines and Resources, RG 21/41/7; Research Council of Alberta, Technical Advisory Committee, Minutes, 21 November 1952, PAA, PP, 1894B; Karl Clark to Ed Nelson, 8 January 1953, UAA, MM, 5563; W. Norrish to John Convey, 13 August 1954, RG 21/41/7.

5. J.E. Sydie to E.C. Manning, 18 August 1953; E.C. Manning to J.E. Sydie, 11 September 1953, both in PP, 1645.

6. Can-Amera Oils to Deputy Minister, Mines and Minerals, 18 May 1954; H.H. Sommerville to E.C. Manning, 25 May 1954, PP, 1645. K.A. Clark to Dr. Grace, 5 November and 30 December 1954; Gordon Taylor and Ivan Casey, 17 January 1955; both in PP, 1643.

7. August W. Giebelhaus, *Business and Government in the Oil Industry: A Case Study of Sun Oil, 1876-1945* (Greenwich, Conn.: JAI Press, 1980), Chapters 4-8; Arthur W. Johnson, *The Challenge of Change: The Sun Oil Company 1945-1977* (Columbus: Ohio State University Press, 1983), 125-61; J.K. Galbraith, *The New Industrial State* (Boston: Houghton Mifflin, 1967), 123-132 et. seq.

8. Great Canadian Oils Sands Brief to Premier Manning, 29 November 1954; Leslie Blackwell to Premier Manning, 3 and 30 November, 16 December 1954, all in PP, 1646; Blackwell to Manning, 30 June 1955, PP, 1745B; W.H. Norrish, memorandum re: Abasand Oils, 23 March 1954; George Hume to Douglas Robinson, 23 February 1956, both in RG 21/46/18; E.C. Manning to L. Blackwell, 7 December 1954, PP, 1646; I.N. Mackinnon to Louis Kjorstand, 2 April 1953; I.N. Mackinnon to E.C. Manning, 4 May 1953, both in PP, 1645; J.E. Oberholtzer to R.A. McMullen, 20 December 1963, PAA, APS Agent-General Records, 526; Larry Pratt, *The Tar Sands: Syncrude and the Politics of Oil* (Edmonton: Hurtig Publishers, 1976), 44 ff; J. Joseph Fitzgerald, *Black Gold With Grit: The Alberta Oil Sands* (Sidney, B.C.: Gray's Publishing Ltd., 1978), 159 ff; Johnson, *The Challenge of Change*, 161.

9. Pratt, *The Tar Sands*; Fitzgerald, *Black Gold With Grit*, 181 ff; G. Bruce Doern and Glen Toner, *The Politics of Energy. The Development and Implementation of the* NEP (Toronto: Methuen, 1984), 134-138. On Alberta's policy see the Appendix explaining Alberta's prorationing system as it applied to oil sands in the essay by an executive with Syncrude Canada Ltd., F.K. Spragins, "Athabasca Tar Sands: Occurrence and Commercial Projects," in *Bitumens, Asphalts and Tar Sands*, edited by G.V. Chilingarian and T.F. Yen (Amsterdam, New York: Elsevier Scientific Pub. Co., 1978), 277 and 278.

10. Economic Council of Canada, *Connections: An Energy Strategy for the Future* (Ottawa: Supply and Services Canada, 1985), 44-45.

11. Carl Berger, *The Sense of Power: Studies in the Ideas of Canadian Imperialism 1867-1914* (Toronto: University of Toronto Press, 1970). Douglas R. Owram, *The Promise of Eden: The Canadian Expansionist Movement and the Idea of the West, 1856-1900* (Toronto: University of Toronto Press, 1980).

12. John Richards and Larry Pratt, *Prairie Capitalism: Power and Influence in the New West* (Toronto: McClelland and Stewart, 1979). Douglas R. Owram, "The Economic Development of Western Canada: An Historical Overview," Economic Council of Canada *Discussion Paper* No. 219, 1982.

13. The document is reprinted in Douglas R. Owram ed., *The Formation of Alberta: A Documentary History* (Calgary: Alberta Records Publication Board, Historical Society of Alberta, 1979), document VI-36.

14. Michael Bliss, *The Discovery of Insulin* (Toronto: McClelland and Stewart, 1982). S.E.D. Shortt, "Banting, Insulin and the Question of Simultaneous Discovery," *Queen's Quarterly* 89, no. 2 (Summer 1982): 260-73. Johnson, *The Challenge of Change*, 127-29. Each of these accounts contrasts with the tendency among writers on the oil sands to assess the claims to discovery of Clark or Ells.

APPENDIX

1. R.J. Giguere, "In-Situ Recovery Process for the Oil Sands of Alberta," Canadian Chemical Engineering Conference, *26th Conference*, October 1976.

2. Karl Clark, "The Athabasca Tar Sands," *Scientific American* 180, no. 5 (May 1949).

3. L.E. Djingheuzian, "Cold-Water Method of Separation of Bitumen from Alberta Bituminous Sands," *Proceedings*, Athabasca Oil Sands Conference, edited by K.A. Clark (Edmonton: Oil Sands Project, Board of Trustees, 1951). S.M. Blair, "Report on the Alberta Bituminous Sands," Alberta *Sessional Paper* no. 49, 1951.

4. Djingheuzian, "Cold-Water Method of Separation."

5. B.D. Sparks and F.W. Meadus, "A Combined Solvent Extraction and Agglomeration Technique for the Recovery of Bitumen from Tar Sand," *Energy Processing Canada* 72, no. 1 (September-October 1979).

6. See Alberta Oil Sands Technology and Research Authority (AOSTRA), *Sixth Annual Report* (Edmonton: Alberta Oil Sands Technology and Research Authority, 1981).

7. W.S. Peterson and P.E. Gishler, "The Fluidized Solids Technique applied to Alberta Oils Sands Problem," in *Proceedings*, Athabasca Oil Sands Conference.

8. Ibid.

9. Alberta Oil Sands Technology and Research Authority (AOSTRA), *Fifth Annual Report and Five Year Review* (Edmonton: Alberta Oil Sands Technology and Research Authority, 1980). See also AOSTRA, *Sixth Annual Report*.

10. W. Taciuk, "Oil Sands Treatment Utilizing the Taciuk Direct Thermal Processor," *Energy Processing Canada* 74, no. 4 (September-October 1981).

11. J.H. Cottrell, "Development of the Anhydrous Process for Oil Sand Extraction," in *The K.A. Clark Volume* (Edmonton: Research Council of Alberta, 1963).

12. J. Joseph Fitzgerald, *Black Gold With Grit: The Alberta Oil Sands* (Sidney, B.C.: Gray's Publishing Ltd., 1978).

13. T.M. Doscher *et al.*, "Steam Drive: A Process for In-Situ Recovery of Oil from the Athabasca Oil Sands," in *The K.A. Clark Volume*.

14. Sidney T. Fisher, "Four Processes for Synthetic Crude," *Energy Processing Canada* 74, no. 1 (September-October 1981).

15. M.L. Natland, "Project Oil Sand," in *The K.A. Clark Volume*.

16. C.M. Davis, "Electrovolatization of Oil In Situ," *Proceedings*, Athabasca Oil Sands Conference.

17. Alfred R.C. Selwyn, "Summary Reports of the Operations of the Geological Corps to 31 December 1881, and to 31 December 1882," Geological Survey of Canada *Report of*

Progress for 1880-81-82 (Montreal: Dawson Brothers, 1883).

18. For detailed references, see the discussion of Ells in Chapter I.

19. Research Council of Alberta, *Tenth Annual Report* (Edmonton: Research Council of Alberta, 1929). For detailed references, see the discussion of Clark's work in Chapter II.

20. For detailed references, see the discussion of Fitzsimmons's work in Chapter III.

21. For detailed references, see the discussion of Ball's work in Chapter III.

22. For detailed references, see the discussion of the federal government's Abasand project in Chapter IV.

23. For detailed references, see the discussion of the Alberta government's Bitumount project in Chapter V.

24. S.M. Blair, "Report on the Alberta Bituminous Sands."

Bibliography

MANUSCRIPT SOURCES

1. *Glenbow-Alberta Institute Archives.*
   Alberta Oil Examiner, 1926. In 1927 became the *Western Oil Examiner,* 1927-51.

2. *Provincial Archives of Alberta.*
   Alberta. Department of Mines and Minerals.
   Alberta. Department of Municipal Affairs.
   Alberta. Department of Provincial Secretary.
   Alberta. Department of Provincial Treasurer.
   Alberta. Department of Public Works.
   Alberta. Department of Public Works, Ministers' File.
   Alberta. Department of Transportation.
   Alberta. Energy Resources Conservation Board.
   Alberta. Premiers' Papers:
      William Aberhart.
      John E. Brownlee.
      Herbert Greenfield.
      Ernest C. Manning.
      R.G. Reid.
      Charles Stewart.
   Karl Adolph Clark.
   Consolidated Mining and Smelting Co.
   Robert Fitzsimmons and International Bitumen Co.
   Charles Knight.
   A.C. Rutherford.
   Achilles Schmid.
   Kenneth Stuart.

3. *Public Archives of Canada.*
   Canada. Department of Energy, Mines and Resources.
   Canada. Department of Indian Affairs and Northern Development.
   Canada. Department of Munitions and Supply.
   S.C. Ells.
   Geological Survey of Canada.
   C.D. Howe.
   Mineral Resources Branch.

Mines Branch.
Earl Smith.

4. *University of Alberta Archives.*
   John A. Allen.
   Sidney R. Blair.
   A.E. Cameron.
   Lloyd Champion.
   Karl A. Clark.
   Department of Mining and Metallurgy.
   Faculty of Engineering.
   William Pearce.
   Presidents' Papers:
      W.A.R. Kerr.
      Henry Marshall Tory.
      Robert C. Wallace.
   Research Council of Alberta.
   Louis A. Romanet.
   Science Association of the University of Alberta.

PUBLISHED SOURCES

Adkins, W.E. "New Plant to Process Athabaska Oil Sands." *Petroleum Engineering* (April 1948): 121-126.

Alberta. Energy and Natural Resources. "Alberta Oil Sands: Facts and Figures." Energy and Natural Resources *Report* 110, 1979.

_____. Energy Resources Conservation Board. "Alberta's Reserves of Crude Oil...." Energy Resources Conservation Board *Report* 81-18, 1981.

_____. Energy Resources Conservation Board. "Crude Bitumen Reserves ... 1980." Energy Resources Conservation *Report* 81-38, June 1981.

Alberta. Department of Education. *Junior Atlas of Alberta.* Cartographically edited by J.C. Muller and L. Wonders. Edmonton: Curriculum Branch, Alberta Education, 1979.

Alberta. Department of the Provincial Treasurer. *Government Estimates,* 1985-86. Edmonton: 1985.

Alberta. Department of the Provincial Treasurer. *Public Accounts of Alberta*. Edmonton: Provincial Auditor, selected years from 1905.

Alberta Oil Sands Technology and Research Authority (AOSTRA). *Fifth Annual Report and Five Year Review*. Edmonton: Alberta Oil Sands Technology and Research Authority, 1980.

_____. *Sixth Annual Report*. Edmonton: Alberta Oil Sands Technology and Research Authority, 1981.

_____. *Ninth Annual Report*. Edmonton: Alberta Oil Sands Technology and Research Authority, 1984.

Alberta. Petroleum and Natural Gas Conservation Board. *Alberta Oil Industry*. N.p. Selected years after 1948.

Alberta. *Royal Commission to Enquire into Matters Connected with Petroleum and Petroleum Products*. Calgary: 1940.

Allan, J.A. "The Mineral Resources of Alberta." Research Council of Alberta *Report* no. 2, 1921.

Allen, A.R. and Sanford, E.R. "The Great Canadian Oil Sands Operation." In *Guide to the Athabasca Oil Sands Area*, 103-121. Research Council of Alberta Information Series, no. 65. 1973.

Armstrong, Christopher. *The Politics of Federalism: Ontario's Relations with the Federal Government 1867-1942*. Toronto: University of Toronto Press, 1981.

Ball, Max. "Development of the Athabaska Oil Sands." *Transactions*, Canadian Institute of Mining and Metallurgy 44 (1941): 58-91.

_____. *This Fascinating Oil Business*. Indianapolis and New York: The Bobbs Merrill Company, 1940.

Barr, John J. *Dynasty: The Rise and Fall of Social Credit in Alberta*. Toronto: McClelland and Stewart, 1974.

Barry, P.S. "The Prolific Pipeline." *Dalhousie Review* 57 (Summer 1977): 205-23.

Bell, Robert. "Report on Part of the Basin of the Athabasca River, North West Territory." Canada. Geological Survey of Canada, *Report of Progress for 1882-84*, 5-35. Montreal: Dawson Brothers, 1884.

Berger, Carl. *The Sense of Power: Studies in the Ideas of Canadian Imperialism 1867-1914*. Toronto: University of Toronto Press, 1970.

Bladen, M.L. "Construction of Railways in Canada, 2, 1885-1931." *Canadian Contributions to Economics* 7 (1934).

Blair, John M. *The Control of Oil*. New York: Vintage Trade Books, 1978.

Blair, S.M. "Report on the Alberta Bituminous Sands." Alberta *Sessional Paper* no. 49, 1951.

Bliss, Michael. *The Discovery of Insulin*. Toronto: McClelland and Stewart, 1982.

Bothwell, Robert and English, John; and Drummond, Ian. *Canada Since 1945: Power and Politics and Provincialism*. Toronto: University of Toronto Press, 1981.

Bothwell, Robert and Kilbourn, William. *C.D. Howe: A Biography*. Toronto: McClelland and Stewart, 1979.

Breen, David H. "Anglo-American Rivalry and the Evolution of Canadian Petroleum Policy to 1930." *Canadian Historical Review* 62 (Summer 1981): 283-303.

Breen, David H., ed. *William Stewart Herron: Father of the Petroleum Industry in Alberta*. Calgary: Alberta Records Publication Board, 1984.

Brown, J.J. *Ideas in Exile. A History of Canadian Invention*. Toronto: McClelland and Stewart, 1967.

Camp, F.W. "Tar Sands." *Encyclopedia of Chemical Technology*. 2nd ed. New York: John Wiley and Sons, 1963.

Canada. Bureau of Mines. "The Canadian Mineral Industry in 1944." Bureau of Mines *Report* no. 815. Ottawa: King's Printer, 1945.

_____. _____. "Drilling and Samples of Bituminous Sands of Northern Alberta ... 1942-1947." Bureau of Mines *Report* no. 826, Ottawa: King's Printer, 1948.

_____. Department of Mines. *Mineral Production Statistics in Canada*. Annual Reports, 1911-1951. Ottawa: King's Printer.

_____. Department of Energy, Mines and Resources. "Oil Sands and Heavy Oils: The Prospects." Department of Energy, Mines and Resources *Report* EP 77-2, 1977.

_____. Department of Indian Affairs and Northern Development. *North of 60. Oil and Gas Statistical Report* no. 3, 1920-81. Ottawa: Supply and Services Canada, 1984.

_____. Mines Branch. "Investigation of Mineral Resources in Canada 1930." Mines Branch *Report* no. 723. Otttawa: King's Printer, 1931.

_____. Mines Branch. "The Canadian Mining Industry in 1935." Mines Branch *Report* no. 773. Ottawa: King's Printer, 1936.

Canadian Petroleum Association. *Statistical Handbook.* Calgary: Canadian Petroleum Association, 1980.

Carrigy, M.A. comp. "Athabasca Oil Sands Bibliography." Edmonton: Research Council of Alberta *Preliminary Report* 63-5, 1965.

_____. "Introduction and General Geology." In *Guide to the Athabasca Oil Sands Area*, 1-14. Edmonton: Research Council of Alberta *Information Series, no.* 65, 1973.

_____. "Mesozoic Geology of the Fort McMurray Area." In *Guide to the Athabasca Oil Sands Area*, 77-101. Edmonton: Research Council of Alberta *Information Series, no.* 65, 1973.

Chalmers, John W., ed. *The Land of Peter Pond.* Edmonton: Boreal Institute of Northern Studies, University of Alberta, Occasional Publication no. 12, 1974.

_____. "A Century of the Fur Trade." In *The Land of Peter Pond.* Edited by John W. Chalmers, 45-62. Edmonton: Boreal Institute of Northern Studies, University of Alberta, Occasional Publication, no. 12, 1974.

Clark, K.A. "Athabasca Oil Sands." *Edmonton Geological Quarterly* 1, nos. 1-2 (1957): 3-5, and 1-5.

_____. "The Athabasca Tar Sands." *Scientific American* 180 (May 1949): 52-55.

_____. "The Separation of Bitumen." *Transactions,*

Canadian Institute of Industrial Metallurgy 32, (1929): 344-59.

_____. "Tar Sands." *Encyclopedia of Chemical Technology.* 1st ed. New York: John Wiley and Sons, 1947.

Clark, K.A., ed. *Proceedings,* Athabasca Oil Sands Conference, September 1951. Edmonton: Board of Trustees, Oil Sands Project, 1951.

Clark, K.A. and Blair, S.M. "The Bituminous Sands of Alberta." Research Council of Alberta *Report No.* 18, 3 vols. Edmonton: Research Council of Alberta, 1927-29.

Clark, K.A. and Pasternack, D.S. "Hot Water Separation of Bitumen from Alberta Bituminous Sands." *Industrial and Engineering Chemistry* 24 (December 1932): 1410-1416.

Comfort, Darlene J. *The Abasand Fiasco: The Rise and Fall of a Brave Oil Sands Extraction Plant.* Edmonton: Friesen Printers, 1980.

_____. "William McMurray: The Man Behind the Fort." *Alberta History* 23, no. 4, (1975): 1-5.

_____. "Tom Draper: Oil Sands Pioneer." *Alberta History* 25, no. 1, (1977): 25-29.

Cottrelle, J.H. "Development of the Anhydrous Process for Oil Sand Extraction." In *The* K.A. *Clark Volume,* 193-206. Edmonton: Research Council of Alberta, 1963.

Corbett, E.A. *Henry Marshall Tory: Beloved Canadian.* Toronto: Ryerson Press, 1954.

Danielson, Albert L. *The Evolution of* OPEC. New York: Harcourt Brace Jovanovich, 1982.

Davis, C.M. "Electrovolatization of Oil In Situ." *Proceedings,* Athabasca Oil Sands Conference. Edited by K.A. Clark, 141-152. Edmonton: Oil Sands Project, Board of Trustees, 1951.

Dawson, George M. "Summary Report." Geological Survey of Canada *Annual Report* 1894 7, 3A-124A. Ottawa: Queen's Printer, 1896.

_____. "Summary Report." Geological Survey of Canada *Annual Report* 1897 10, 3A-156A. Ottawa: Queen's Printer, 1899.

_____. "Summary Report." Geological Survey of Canada *Annual Report* 1898 11, 3A-208A. Ottawa: Queen's Printer, 1901.

Demaison, G.J. "Tar Sands and Supergiant Oil Fields." *Canadian Institute of Mining* 17 (Annual), 1977: 9-16.

Diubaldo, Richard J. "The Canol Project in Canadian American Relations." Canadian Historical Association, *Historical Papers*, (1977): 178-95.

Djingheuzian, L.E. "Cold-Water Method of Separation of Bitumen from Alberta Bituminous Sand." *Proceedings*, Athabasca Oil Sands Conference. Edited by K.A. Clark. 185-199. Edmonton: Oil Sands Project, Board of Trustees, 1951.

Doern, G. Bruce and Glen Toner. *The Politics of Energy: The Development and Implementation of the* NEP. Toronto: Methuen, 1984.

Doscher, T.M. *et al.* "Steam Drive: A process for In-Situ Recovery of Oil from the Athabasca Oil Sands." In *The* K.A. *Clark Volume*, 123-141. Edmonton: Research Council of Alberta, 1963.

Dranchuk, P.M. "Exploitation of the Oil Sands of Alberta." *Proceedings, Ninth Annual Conference,* Canadian Technical Asphalt Association 19 (1974): 171-185.

Economic Council of Canada. *Connections: An Energy Strategy for the Future.* Ottawa: Supply and Services Canada, 1985.

Ells, S.C. "Preliminary Report on Bituminous Sands." Canada. Mines Branch *Report* no. 281. Ottawa: King's Printer, 1914.

_____. "The Bituminous Sands of Northern Alberta." Canada Mines Branch *Report* no. 626. Ottawa: King's Printer, 1924.

_____. "Bituminous Sands of Northern Alberta." Canada. Mines Branch *Report* no. 632. Ottawa: King's Printer, 1926.

_____. "Uses of Alberta Bituminous Sands for Surfacing Roads." Canada. Mines Branch *Report* no. 684. Ottawa: King's Printer, 1927.

_____. "Recollections of the Development of the Athabasca Oil Sands." Canada. Mines Branch *Information Circular* 139. Ottawa: Queen's Printer, 1962.

Finnie, Richard. *Canol. The Sub-Arctic Pipeline and Refinery Project Constructed by Bechtel-Price-Callaghan for the Corps of Engineers United States Army, 1942-44.* San Francisco: Produced for Bechtel-Price-Callaghan by Ryder and Ingram, Publishers, 1945.

Fisher, Sidney T. "Four Processes for Synthetic Crude." *Energy Processing Canada* 74, no. 1 (September-October 1981): 20-26.

Fitzgerald, J. Joseph. *Black Gold With Grit: The Alberta Oil Sands.* Sidney, B.C.: Gray's Publishing Ltd., 1978.

Freeman, J.M. *Biggest Sellout in History; Foreign Ownership of Alberta's Oil and Gas Industry and the Oil Sands.* Edmonton: Alberta's New Democratic Party, 1966.

Fowke, V.C. *The National Policy and the Wheat Economy.* Toronto: University of Toronto Press, 1957.

Fumoleau, Rene. *As Long As This Land Shall Last: A History of Treaties 8 and 11 (1870-1939).* Toronto: McClelland and Stewart, 1973.

Galbraith, A.D. "Oil from the Oil Sands." *Proceedings*, Ninth Annual Conference, Canadian Technical Asphalt Association 19 (1974): 187-93.

Galbraith, J.K. *The New Industrial State.* Boston: Houghton Mifflin, 1967.

Gander, James E. and Belaire, Fred. W. "Energy Futures for Canada." Canada. Energy, Mines and Resources. *Report* EP 78-1. Ottawa: Supply and Services, 1978.

Gates, Charles M., ed. *Five Fur Traders of the Northwest.* Minneapolis: University of Minnesota Press, 1933.

Geological Survey of Canada. *Report of Progress for* 1880-81-82. Montreal: Dawson Brothers, 1883.

Giebelhaus, August W. *Business and Government in the Oil Industry: A Case Study of Sun Oil, 1876-1945.* Greenwich, Conn.: JAI Press, 1980.

Giguere, R.J. "An In-Situ Recovery Process for the Oil Sands of Alberta." Canadian Chemical Engineering Conference, 26th *Conference*, October 1976.

Granatstein, J.L. *Canada's War: The Politics of the Mackenzie King Government, 1939-1945*. Toronto: Oxford University Press, 1975.

_____. *A Man of Influence: Norman A. Robertson and Canadian Statecraft 1929-68*. Ottawa: Deneau Publishers, 1981.

Gray, Earle. *Impact of Oil: The Development of Canada's Oil Resources*. Toronto: Ryerson Press Maclean-Hunter, 1969.

_____. *The Great Canadian Oil Patch*. Toronto: Maclean-Hunter, 1970.

_____. *Wildcatters: The Story of Pacific Petroleums and Westcoast Transmission*. Toronto: McClelland and Stewart, 1982.

Griffin, Harold. *Alaska and the Canadian Northwest*. New York: W.W. Norton, 1944.

Hanson, Eric. *Dynamic Decade: The Evolution and Effects of the Oil Industry in Alberta*. Toronto: McClelland and Stewart, 1958.

Harcourt, George. "Economic Resources of Alberta." In *Canada and Its Provinces. A History of the Canadian People and Their Institutions. By One Hundred Associates*. 20. Edited by A. Shortt and A.G. Doughty, 583-601. Edinburgh edition. Toronto: Edinburgh University Press for the Publishers Association of Canada, 1914.

Ingall, Elfric Drew. "Division of Mineral Statistics and Mines Annual Report for 1890." Geological Survey of Canada *Annual Report for 1890-1891* 5, Part 2, 1S-200S. Ottawa: Queen's Printer, 1893.

Irwin, J.L. "Oil Sands Symposium." *Petroleum Times*, 5 October 1951: 884.

Jacoby, Neil H. *Multinational Oil: A Study in Industrial Dynamics*. New York: Macmillan, 1974.

Johnson, Arthur M. *The Challenge of Change: The Sun Oil Company 1945-1977*. Columbus: Ohio State University Press, 1983.

Kennedy, J. de N. History of the Department of Munitions and Supply, 1. Ottawa: E. Cloutier, 1950.

Lamb, W. Kaye, ed. Journals and Letters of Alexander Mackenzie. Hakluyt Society Extra Series no. 4. Toronto: Macmillan of Canada, 1970.

Lithwick, N.H. Canada's Science Policy and the Economy. Toronto: Methuen, 1969.

MacGregor, James G. A History of Alberta. Edmonton: Hurtig Publishers, 1972.

Mackinnon, C.S. "Some Logistics of Portage La Loche." Prairie Forum 5, no. 1 (1980): 51-65.

Macoun, John. "Geological and Topological Notes by Professor Macoun, on the Lower Peace and Athabasca Rivers." Appendix I, 87-95. In Geological Survey of Canada Report of Progress for 1875-76. Montreal: by Authority of Parliament, 1877.

Macpherson, C.B. Democracy in Alberta: Social Credit and the Party System. Toronto: University of Toronto Press, 1953.

Mallory, J.R. Social Credit and the Federal Power in Canada. Toronto: University of Toronto Press, 1954.

McDougall, John N. Fuels and the National Policy. Toronto: Butterworths, 1982.

de Mille, George. Oil in Canada West: The Early Years. Calgary: Northwest Printing and Lithographing, 1970.

Morton, W.L. The Critical Years: British North America 1857-1873. Toronto: McClelland and Stewart, 1964.

Natland, M.L. "Project Oilsand." In The K.A. Clark Volume, 143-157. Edmonton: Research Council of Alberta, 1963.

Nelles, H.V. The Politics of Development: Forests, Mines and Hydro-Electric Power in Ontario, 1849-1941. Toronto: Macmillan of Canada, 1974.

de Never, Noel. "Tar Sands and Oil Shales." Scientific American 214, (February 1966): 21-29.

Norris, A.W. "Paleozoic (Devonian) Geology of Northeastern Alberta and Northwestern Saskatchewan." In *Guide to the Athabasca Oil Sands Area*, 15-76. Edmonton: Research Council of Alberta Information Series, no. 65, 1973.

Owram, Douglas R., ed. *The Formation of Alberta: A Documentary History*. Calgary: Alberta Records Publication Board, Historical Society of Alberta, 1979.

_____. *The Promise of Eden: The Canadian Expansionist Movement and the Idea of the West, 1856-1900*. Toronto: University of Toronto Press, 1980.

_____. "The Economic Development of Western Canada: An Historical Overview." Economic Council of Canada *Discussion Paper* no. 219, 1982.

Parker, G. "Alberta Bituminous Sands for Rural Roads." Canada. Mines Branch. *Report for* 1918. *Sessional Paper* no. 26a, 1920.

Parker, J.M. "The Long Technological Search." In *The Land of Peter Pond*. Edited by J.W. Chalmers, 109-119. Edmonton: Boreal Institute of Northern Studies, University of Alberta, Occasional Publication no. 12, 1974.

_____. "History of the Athabasca Oil Sands Region 1890 to 1960s, 2, Oral History." Alberta Oil Sands Environmental Program Project HS 10.1, 1980.

Parker, J.M. and Tingley, K.W. "History of the Athabasca Oil Sands Region 1890-1960s, 1." Alberta Oil Sands Environmental Research Program Project HS 10.2, 1980.

Peterson, W.S. and Gishler, P.E. "The Fluidized Solids Technique Applied to Alberta Oil Sands Problem." *Proceedings*, Athabasca Oil Sands Conference, 207-236. Edmonton: Oil Sands Project, Board of Trustees, 1951.

Pow, J.R.; Fairbanks, G.H.; and Zamora, W.J. "Descriptions and Reserves Estimates of the Oil Sands of Alberta." Edmonton: Research Council of Alberta, *Information Report* 45, 1963.

Pratt, Larry. *The Tar Sands: Syncrude and the Politics of Oil*. Edmonton: Hurtig Publishers, 1976.

Pratt, Larry and Stevenson, Garth, eds. *Western Separatism: Myths, Realities and Dangers*. Edmonton: Hurtig Publishers, 1981.

Pratt, Wallace. "Max W. Ball." *Bulletin*, American Association of Petroleum Geologists 39, no. 5 (May 1955): 775-80.

Procter, R.M.; Taylor, Gordon C.; and Wade, John A. "Oil and Natural Gas Resources of Canada 1983." Geological Survey of Canada *Paper* 83-31. Ottawa: Geological Survey, 1983.

Rasporich, A.W., ed. *The Making of the Modern West. Western Canada since* 1945. Calgary: University of Calgary Press, 1984.

Redford, D. and Winstock, A. "Oil Sands of Canada and Venezuela." *Canadian Institute of Mining* 17 (Special Volume), 1977.

Research Council of Alberta. *Annual Reports*, 1920-1933. Edmonton: Research Council of Alberta, 1921-34.

Richards, John and Pratt, Larry. *Prairie Capitalism: Power and Influence in the New West*. Toronto: McClelland and Stewart, 1979.

Richardson, John. *Arctic Searching Expedition: A Journal of a Boat-Voyage through Rupert's Land and the Arctic Sea*, 1. London: Longman, Brown, Green and Longmans, 1851.

Rosewarne, P.V. and Swinnerton, A.A. "Report of Laboratory Investigations on the Cold Water Separation of Bitumen from Alberta Bituminous Sands." Canada Bureau of Mines *Report* No. 90. Ottawa: King's Printer, 1948.

Ross, Victor. *Petroleum in Canada*. Toronto: Southam Press, 1917.

Selwyn, Alfred R.C. "Summary Reports of the Operations of the Geological Corps to 31st December, 1881, and to 31st December, 1882." Geological Survey of Canada *Report of Progress for* 1880-81-82, 1-29. Montreal: Dawson Brothers, 1883.

Shortt, S.E.D. "Banting, Insulin and Question of Simultaneous Discovery." *Queen's Quarterly* 89, no. 2 (1982): 260-73.

Smith, Philip. *The Treasure-Seekers: The Story of Home Oil*. Toronto: Macmillan of Canada, 1978.

Sparks, B.D. and Meadus, F.W. "A Combined Solvent Extraction and Agglomeration Technique for the Recovery of Bitumen from Tar Sand." *Energy Processing Canada* 72, no. 1 (September-October 1979): 55-61.

Spragins, F.K. "Athabasca Tar Sands: Occurrence and Commercial Projects." In *Bitumens, Asphalts and Tar Sands,* edited by G.V. Chilingarian and T.F. Yen, 93-121. Amsterdam, New York: Elsevier Scientific Pub. Co., 1978.

Staff, Oil Sands Project |W.E. Adkins|. "Report to the Board of Trustees on the Oil Sands Project from Inception to December 31, 1948." Alberta. *Sessional Paper* no. 53, 1949.

_____. "Report to the Board of Trustees on the Alberta Government Oil Sands Project from January 1, 1949 to December 31, 1949." Alberta *Sessional Paper* no. 55, 1950.

Stanley, G.F.G. *The Birth of Western Canada: A History of the Riel Rebellion,* London: Longmans, Green, 1936; Toronto: University of Toronto Press, reprint, 1961.

Statistics Canada. "Crude Petroleum and Natural Gas Industry." *Annual* 26-213 |selected years|.

Stewart, G.A. "Geological Controls on the Distribution of Athabasca Oil Sands Reserves." Research Council of Alberta *Information Report* no. 45, 1963.

Taciuk, W. "Oil Sands Treatment Utilizing the Taciuk Direct Thermal Processor." *Energy Processing Canada* 74, no. 4 (September-October 1981): 27-30.

Thomas, L.H., ed. *William Aberhart and Social Credit in Alberta.* Toronto: Copp Clark, 1977.

Towson, Donald. "Tar Sands." *Encyclopedia of Chemical Technology.* 3rd ed. New York: John Wiley and Sons, 1978.

Traves, Tom. *The State and Enterprise: Canadian Manufacturers and the Federal Government, 1917-1931.* Toronto: University of Toronto Press, 1979.

Vietor, Richard H.L. "The Synthetic Liquid Fuels Program: Energy Politics in the Truman Era." *Business History Review* 54, no. 1 (Spring 1980): 1-34.

Waite, P.B. *Canada 1874-1896: Arduous Destiny*. Toronto: McClelland and Stewart, 1971.

Watkins, Ernest. *The Golden Province: Political Alberta*. Calgary: Sandstone Publishing Ltd., 1980.

Weir, Thomas R., and Matthews, Geoffrey J. *Atlas of the Prairie Provinces*. Toronto: Oxford University Press, 1971.

Wonders, W.C. "Wood and Water, Land and Oil." In *The Land of Peter Pond*. Edited by John W. Chalmers, 9-22. Edmonton: Boreal Institute of Northern Studies, University of Alberta, Occasional Publication no. 12, 1974.

Zaslow, Morris, *The Opening at the Canadian North, 1870-1914*. Toronto: McClelland and Stewart, 1971.

_____. *Reading the Rocks: The Story of the Geological Survey of Canada 1842-1972*. Toronto: Macmillan of Canada in Association with the Department of Energy, Mines and Resources, and Information Canada, 1975.

UNPUBLISHED SOURCES

Alberta Culture. Heritage Sites Service. "Athabasca Oil Sands Development, Bitumount Alberta." 1974.

Baillie, R.A. and Mertes, T.S. "Development and Production of Oil from the Athabasca Tar Sands." Marcus Hook, Pa.: Sun Oil Co., 1968 (Suncor Inc. Library, Calgary).

Ball, Max. "Oil In the Oil Sands of Alberta Canada." Paper presented at the annual meeting of the American Institute of Mining and Metallurgical Engineers, New York, February 1937. (University of Alberta Archives, L.A. Romanet Papers, vol. 7).

Canadian Energy Research Institute. "The Wolf Lake Project." Client Study. Calgary, 1984.

Clark, K.A. "Hot Water Method for Recovering Bitumen from Bituminous Sand." Paper written at Sullivan concentrator, Chapman Camp, British Columbia for Consolidated Mining and Smelting Co. 1939. In *Papers Relating to Alberta Bituminous*

Sand 1930-50. By Karl A. Clark. (Bound in the Library of the University of Alberta.)

Ells, S.C. "Notes on certain aspects of the Proposed Commercial development of the deposits of Bituminous Sands in the Province of Alberta, Canada." 2 vols., typescript. Canada. Mines Branch, 1917. (University of Alberta Archives, Department of Mining Engineering Papers.)

Howes, D.A. "The Alberta Bituminous Sands." Alberta. Department of Transportation, 1951. (Provincial Archives of Alberta.)

Korvemaker, E. Frank. "Bitumount, Alberta: A Preliminary Historical Report." Edmonton: Provincial Museum and Archives of Alberta, 1973.

Nicks, John S. "Bitumount: A Preliminary Assessment of Industrial Archaeological Potential and Requirements." Alberta Culture, Historic Sites Service, October 1981.

Riddell, K.M. "Origins and Evolution of the Scientific and Industrial Research Council of Alberta, 1913-1930." Bachelor's Honours Essay, University of Alberta, Department of History, 1977.

Riddell Aytenfisu, K.M. "The University of Alberta: Objectives, Structure and Role in the Community, 1908-1928." Master's thesis, University of Alberta, 1982.

Smith, A.J. and McClave, J.M. "Commercial Utilization of the Oil Sands of Northern Alberta." Paper presented at the 18th annual western meeting of the Canadian Institute of Mining and Metallurgy, October 1936. (University of Alberta Archives, LA. Romanet Papers, vol. 7.)

Index

low levels in, 133, 134
    well (first), 21
Athabasca Landing, 20
Athabasca Oil and Asphalt Co., 60
Athabasca Oil Sands Conference, 143, 147
Athabasca-Spokane Co., 61

B

Badura, Frank, 78
Badura,Marie, 78
Ball, Max, 54, 59, 60, 69, 71, 75, 76, 77, 80, 85, 86, 88, 89, 90, 91, 93, 94, 95, 97, 101,
    102, 105, 106, 107, 108, 109, 111, 162, 165, 202
Ball mill, 175
Banting, Sir F.G., 164
Bechtel Co., 154
Becker, Charles, 82
Bell, Robert, 15, 16, 17, 18, 19, 20, 21, 26, 30, 159, 160, 163
Best, C.H., 164
Bitumen, drilling for. *See* Drilling
    experiments, 60
    lubricants, 49
    motor fuels, 45, 46, 49, 199
    petroleum products, 24, 38, 41, 46, 51, 58, 125
    properties of, 5, 6, 24, 33, 53-54
    road surfacing (paving), 23, 24, 26, 27, 38, 41, 46-48, 51, 58, 59, 60, 62, 63,
        64, 72, 99
    roofing, 72, 75, 199
    sidewalk paving, 47, 64. *See also* Draper, Tom
    separation of, 49. *See* Separation
    uses of, 3, 8, 12, 16, 24, 26, 27, 38, 41, 45, 46, 47, 48, 49, 51, 56, 59-60, 62, 63,
        162, 199. *See also* Athabasca, oil sands
Bituminous Sands Advisory Committee, 51, 88
Bituminous Sands of Alberta, Research Council Report No. 18, 40-43, 45
Bituminous Sands Extraction Co., 68
Bituminous Sands Permit No. 1, 87
Bitumount
    area of, 21, 71
    Board of Trustees, 126, 130, 134, 139, 142
    construction of, 123
    costs, 128, 129, 131, 133, 134, 136, 137, 138, 139, 147
    explosions at separation plant, 1949, 135-37
    fire at, 130
    historic resource, as, 3
    leases, 129, 130
    management, 134
    mining at, 137
    Oil Sands Conference visit to, 143
    plant, 3, 4, 54, 80, 81, 82, 83, 119, 123, 124, 127, 128, 129, 131, 134, 136, 139,
        140, 141, 142, 145, 147, 153, 158, 163, 165, 199, 206-10, 211

operations, (1949), 137
problems, 131, 134
    insufficient diluent, 134
    insufficient water supply, 131, 134, 136, 137
    labour, 134-35
    operating, 134, 207
production, 1948, 131
production, 1949, 137-39
processes, 5
processes at Alberta Government plant, 206-10
processes at International Bitumen, 199-201
processing of Bitumount Sand at Waterways, 198
recovery percentage, 138
refinery, 76, 80, 134, 135, 137-38
report on Bitumount, sessional paper, 1948, 130-32
safety record of, 136, 137
sands, 75
separation plant, 54, 71, 73, 76, 77, 78, 80, 104, 131, 134, 135, 136, 137, 138,
    141, 191, 199-201, 206-10
site, 126, 156
visit of MLAs to, 136-37
wage levels, 82, 131, 132, 134-35
Blair, Sidney, 31, 40, 41, 42, 43, 44, 49, 57, 81, 105, 127, 128, 129, 139, 140, 141, 143,
    151, 153, 156, 163, 165, 167, 191, 210
    Blair's "Report on the Alberta Bituminous Sands," 141-43, 144, 145, 146
Blaylock, S.G., 89, 100, 103, 106, 109
Bliss, Michael, 164
Boomer, E.H., 104
Boot, 194, 195
Born, Sidney, 134
Born Engineering Co., 81, 127, 130, 206
Bosworth, T.O., 61
Brandon, Manitoba, 81
British Admiralty, 22
British North America Act, 124
British Petroleum, 144, 188. See Anglo-Iranian Oils
Brownlee, John E., 50, 52
Bucket elevator, 192, 194, 195

C

Calgary, 22, 68, 85, 153
California, 26, 28, 33, 35, 183
    oil shales, 38
Calvan, 153
Campbell, W.P., 89, 90
Camrose, 63, 64
Camsell, Charles, 51, 114
Can-Amera Oil Sands, 153
Canada

268

D

Dawson Creek, 100
Dehydrator, 53, 72, 198
Denver, 75, 85, 88
Diluent, 90, 107, 111, 112, 114, 115, 116, 131, 137, 171, 174, 175, 176, 180, 204, 205, 208, 211
    kerosene, 90, 174, 207, 210
    naphtha, 90, 173, 174, 204, 210
    mixer, 89, 90, 202
    recovery, 114, 131, 172, 173, 175, 180, 204
    Dilution centrifuging, 198, 199, 205. *See also* Separation
    Direct Thermal Processor, 179
    Distillation processes, 40, 63, 65, 176-79. *See* Separation; Extraction
    Diver, D., 64
    Diver Extraction process, 64, 65
Djingheuzian, L.E.
    cold water separation process, 205. *See* Cold Water Separation; Extraction
Dome, 156
Dominion Government. *See* Canada, federal government
Domtar, 81. *See* Adkins, W.E.
Donald, J.R., 103
Dragline, 90, 200. *See* Strip mining equipment
Draper, Tom, 47, 62-64, 68, 69
    road paving jobs, 63-64
Drilling 20, 21, 30, 46, 59, 60, 62, 65, 71, 152, 182, 187
    core drilling, 20, 105, 111, 152
Drum, 194. *See also* Conditioning drum
Dunvegan Railway Yards, Edmonton. *See* Separation Plants

E

Economic Council of Canada, 156
Embarras River, 14
Edinburgh University, 16
Edmonton, 22, 31, 33, 46, 49, 50, 61, 77, 81, 82, 83, 84, 88, 93, 103, 134, 141, 143, 145, 154
    City Engineer, 73. *See* Haddow, A.W.
    City Engineering Department, 23, 64, 73
    paving, 23, 47
    scientists in, 28
Edmonton Exhibition, 63
Edmonton-St. Albert Trail, 47
Edwards, W.M., 33
Ells, R.W., 22
Ells, Sidney C., 22-29, 30, 33, 36, 38, 48, 51, 56, 62, 68, 71, 81, 84, 87, 104, 110, 159, 160, 164, 191
    Ells's Report on oil sands in Alberta, 1917, 24, 26-27, 31
Emulsifier, 52, 53, 72

271

Grinding flotation unit, 115
Gulf Canada, 155, 156

H

Haanel, B.F., 111, 112, 114
Haanel, Dr. Eugene, 24, 27, 28, 29
Haddow, A.W., 73
Hamar, W.A., 24
Hamilton, P.D., 115, 116, 117
Hammerstein, Von, Alfred, 21, 30, 60, 61, 182
Heath Robinson, 84
Heating Tank, 197
Hinton, William P., 54, 69
Hoffman, Christian G., 18, 19, 21, 26, 30, 191
Hooke, A.J., 145
Hopper. *See* Feed hopper; Strip mining equipment
Horse River, 60, 87,
    oil sands of, 90
    plant, 90, 124, 129, 191, 202, 203
Horse River Reserve, 23, 88
Horizontal rotating processor, 179
Hot Water Extraction. *See* Extraction
Hot Water Separation, 18, 26, 38, 41, 42, 87, 91, 106, 119, 137, 141, 142, 143, 160, 162, 164. *See also* Extraction; Separation plants
Hot Water Washing, 42, 68
Hot Water Washing and Silicate of Soda Mixing Process, 40, 42, 43, 44
Howe, C.D., 92, 101, 103, 109, 110, 111, 117, 118, 119, 125
Howes, D.A., 145, 146, 147, 151, 156, 163
    Howes Report, 145-47, 156
Hudson's Bay Company, 12, 18
Hudson's Bay Oil and Gas, 155
Hume, George, 102, 103, 104, 114
Hydro-cracking, 6, 179

I

Imperial Oil, 61, 76, 91, 154
Indians, 11, 18-19
In-situ combustion (fireflooding), 186-88. *See* Extraction
In-situ extraction, 168, 182-89, 211. *See* Extraction
In-situ processing, 8
In-situ production, 168
In-situ recovery, 8, 129, 143, 168, 169, 188
Insulin, 164
International Bitumen Co., 59, 62, 69, 70, 72, 75, 76, 77, 78, 80, 81, 82, 83, 84, 85, 88, 94, 99, 124, 162, 163, 164, 165, 191, 198, 199, 201, 211
    Revenues of (1938), 82
    Wages at, 82
Irwin, John, 112, 118

274

Venezuela, 98, 146. *See also* Orinoco River
Vibrating screen. *See* Screen
Victoria's Settlement, 20
      well at, 21

W

Wabasca, tar sands, 20
Wage levels, 82, 105, 114, 131, 132, 133, 134, 135
Wainwright, 68
Wales, 68. *See* Lindsay, Major-General Bertram
Wallace, Robert C., 50, 51, 69
Wartime Oils, 110, 117, 119
Water extraction process. *See* Extraction
Water flotation technique. *See* Flotation
Waterman,Isaac, 18
Waterways
      plant, 27, 31, 52, 53, 55, 56, 75, 88, 164, 191, 195-99, 202, 204, 206, 210
      extraction operations at, 195-99
      processing of sands from Bitumount at, 198
Webster, George Boyd, 112, 114, 118, 119
Wellington Street, Ottawa, 63
Wetaskiwin, 68
Wirtz, 82
Wolf Lake, 186, 188

This book is one of a series published jointly by Alberta Culture and the Canadian Plains Research Center. Their common objective is to bring to public notice the diverse aspects of Alberta history as evidenced in Alberta's historical resources. Each is a result of scholarly research by a specialist who has approached the significant historical topic from the vantage point of the associated historic sites.